Praise for
One-Cow Revolution

"Now you can enjoy the deeply practical wisdom of favorite home-steading gurus Shawn and Beth Dougherty in the comfort of your favorite reading chair with this fabulous manual about all things family milk cow. Writing in the same lively style as they speak, Shawn and Beth coach you through the problems you might encounter and offer experience-tested solutions. Even though this material is familiar to me, I couldn't stop reading *One-Cow Revolution*. This is a wear-it-out book; you might want two just to be on the safe side."

—**Joel Salatin**, cofounder, Polyface Farm;
author of *Everything I Want to Do Is Illegal*

"One of the hardest parts about keeping a homestead dairy cow is finding a mentor you can trust to teach you the straightforward 'need to know' information. In this book, Shawn and Beth bring together both their knowledge and decades of hands-on experience for the dairy cow owner. It's as if they are your own personal mentors, sitting with you at your kitchen table. From forage and health to the great grain debate, they tackle it all. This is the homestead dairy cow book every homesteader needs!"

—**Amy Fewell**, founder, Homesteaders of America

"What an inspiring book! As a farmer and homesteader for almost forty years who has shared what I've learned with many others, I find *One-Cow Revolution* to be eminently practical and very encouraging. Shawn and Beth Dougherty exemplify the persistence and tenacity homesteaders must have, cheerfully maintaining a sense of purpose and vision. They don't just tell us what to do but eloquently and articulately inspire us about why we need to do it—for the love of our families, our neighbors, our animals, and the land. They make a compelling case for why we need a One-Cow Revolution!"

—**Charles "Butch" Tindell**, The Ploughshare Institute

"Having a family cow not only allows families to achieve food independence but also vibrant health. Shawn and Beth tell you how to make the family cow a part of your family life."

—**Sally Fallon Morell**, president,
The Weston A. Price Foundation

"Shawn and Beth Dougherty's *One-Cow Revolution* has all the information the beginning homesteader needs to venture into raising a family dairy cow. Having spent all my life around the gentle bovines, the dynamics and considerations for a one-cow situation are quite different than for our fifty-cow herd. The Doughertys clearly show the way—from buying the cow to getting her bred, feeding and milking her, keeping her healthy, and other useful and needed advice. They answer all the questions you were afraid to ask."

—**David Kline**, editor, *Farming Magazine*

"Dairy cows helped create and sustain many low-input Appalachian farms fifty years ago, ours included, and Shawn and Beth do an excellent job of explaining why and how this works. Although they acknowledge that 'the only things you really know are the ones you experience yourself,' their book offers the inspiration for your family to make a life-changing commitment. You'll gain the confidence you need reading the authors' personal anecdotes and stories from other homesteaders on the wonders of improving the health of your land and yourselves with a milk cow. 'The cows teach you, the grass teaches you . . . and observation and reason will teach you.'"

—**Jeff Poppen**, author of *Barefoot Biodynamics*
and *The Barefoot Farmer*

"In many times and places, dairy cattle have been the beating heart of healthy, productive, and sustainable local food systems. In this delightful compendium of light-touch land wisdom, Shawn and Beth Dougherty not only show us how to build such systems today, but also why we need to."

—**Chris Smaje**, author of *Finding Lights in a Dark Age*
and *A Small Farm Future*

"Since moving to our homestead many years ago, my family and I have often wondered how our ancestors used to feed their animals without the aid of store-bought feed and other off-farm inputs. Now I know: it was the family cow. *One-Cow Revolution* by Shawn and Beth Dougherty is a remarkable read, beautifully written and accessible. It reconstitutes knowledge from previous generations for those of us seeking a simpler, more sustainable life. We have two bookshelves in our homesteading library: books that look pretty, and books we reference over and over. *One-Cow Revolution* will be on the latter."

—**Rory Groves**, author of *Durable Trades*

One-Cow
Revolution

Also by Shawn and Beth Dougherty

The Independent Farmstead:
Growing Soil, Biodiversity, and
Nutrient-Dense Food with Grassfed Animals
and Intensive Pasture Management

One-Cow Revolution

Achieving Food Independence with a Grass-Fed Family Cow

Shawn and Beth Dougherty

CHELSEA GREEN PUBLISHING
White River Junction, Vermont
London, UK

First published in 2025 by Chelsea Green Publishing | PO Box 4529 |
    White River Junction, VT 05001 | West Wing, Somerset House, Strand |
    London, WC2R 1LA, UK | www.chelseagreen.com
A Division of Rizzoli International Publications, Inc. | 49 West 27th Street |
    New York, NY 10001 | www.rizzoliusa.com

Publisher: Charles Miers
Deputy Publisher: Matthew Derr
Project Manager: Natalie Wallace
Acquiring Editor: Fern Marshall Bradley
Developmental Editor: Will Solomon
Copy Editor: Deborah Heimann
Proofreader: Nancy Crompton
Indexer: Shana Milkie
Designer: Abrah Griggs

ISBN 978-1-64502-312-8 (paperback) | ISBN 978-1-64502-313-5 (ebook)
Library of Congress Control Number: 2025032143 (print) | 2025032144 (ebook)

Our Commitment to Green Publishing
Chelsea Green sees publishing as a tool for cultural change and ecological stewardship. We strive
to align our book manufacturing practices with our editorial mission and to reduce the impact
of our business enterprise in the environment. We print our books using vegetable-based inks
whenever possible. This book may cost slightly more because it was printed on paper that contains recycled fiber, and we hope you'll agree that it's worth it. *One-Cow Revolution* was printed
on paper supplied by Marquis that is made of recycled materials and other controlled sources.

Authorized EU representative for product safety and compliance
Mondadori Libri S.p.A. | www.mondadori.it
via Gian Battista Vico 42 | Milan, Italy 20123

Printed in Canada.
10 9 8 7 6 5 4 3 2 1   25 26 27 28 29

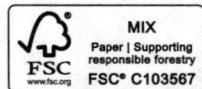

MIX
Paper | Supporting
responsible forestry
FSC
www.fsc.org
FSC® C103567

For Barry

Contents

Introduction

In February of 1986, just two months married, we built a chicken house and planted potatoes on the left bank of the Trinity River in Dallas County, Texas. We had demi-official permission to maintain some kind of low-profile presence on land that was not ours. For the next year, we ate eggs and potatoes and asked ourselves questions about the conditions for human thriving. Mostly we ate potatoes; the chickens disappeared in pairs, generally on Saturday nights, and it was hard to tell whether we felt more irritation or kinship for the human fox who was so careful to leave the door hasped behind him.

Fast-forward ten years—years during which we homesteaded under conditions ranging from a one-room apartment to graduate student housing to a cattle ranch in western Oklahoma—to find us standing on a steep, trash-covered hillside above the Ohio River, wondering what on earth we were doing there. The land had been mined, logged, stripped of its topsoil, and used as a dump. The century-old buildings were in an advanced state of disrepair. In the intervening years we had learned to grow a lot more than potatoes, and we knew a bit about beef cattle, pigs, chickens, and goats, too. But we were still wondering about the conditions that governed real, durable human thriving, and we weren't at all sure we could find answers on land like this.

What we were looking for was something our grandparents, all of whom were born before the dawn of the twentieth century and had farmed with very limited resources through two world wars, could have explained to us if they hadn't died before we knew what we needed to ask. In this age, when every ounce of food, every drop of water, every tool and textile, even every bit of information we receive comes from somewhere else, was made by someone else, we wanted to know: What are the origins of food, health, and beauty? Where do they come from? And how can they be made to keep coming? We were looking for First Causes, and the answers, when they arrived, came piece by piece, dripping out from between the layers of shale on that steep hillside, catching on the legs of our blue jeans as we forced our way through head-high briars, sprouting from moss-covered logs in the woods.

First Causes

Life—pretty much all life—begins as green leaves. The sun shines and by the process of photosynthesis leaves turn light and water and air into plant material, which includes sugars, starches, some proteins and fats, and a lot of micronutrients—but mostly, and by a huge margin, fiber. Fiber makes up as much as 90 percent of the plant material in the world—which means it makes up most of the world's biological energy.

Things that are not plants live by consuming plants, or by consuming creatures that consume plants. And while the plant sugars, starches, fats, and proteins may be used to build and power animal bodies, cellulose, it turns out, is indigestible to macrofauna (large animal life). It's nutritionally inaccessible. Think about that: Nine-tenths of the biological energy of the planet is not—at least, not directly—available to animal life. Fortunately, there's a class of animals that, partnering with fiber-digesting microbes, makes it their whole business to turn cellulose into high-quality and digestible proteins and fats and sugars. These are *ruminants*, grass and forage-eating animals like deer, buffalo, sheep, goats, and cows.

Standing on our weed-choked and briar-infested hillside in Ohio, we were on the cusp of a relationship with this natural ecosystem that was going to teach us where food comes from, what are the conditions for its continuance, and how we might piggyback on our land's natural gifts to increase its productivity and include food for humans—without upsetting the balance. Grass—or rather, all the species belonging to the native plant communities our land was already growing—would show itself to be the source of fertility and food for the whole living

landscape; and the key that would unlock them for our use was the homely family dairy cow.

Grass, in the guts of a cow, becomes meat, manure, and milk; produced all day, every day, all year long, this is how solar energy is continually converted into nutrient forms that will feed not only all the members of the homestead community but the soil as well. Milking a dairy cow—showing up daily or twice daily to see to her needs and receive her lactic blessings—lets us power the whole farm almost entirely from grass.

That steep, rocky hillside had a lot to teach us, and it turned out to be a very good teacher. More congenial land would have given its gifts indulgently, sparing the rod and spoiling the children. But we, unable to exact submission from our inhospitable acres, baffled in our attempts to remake it in the image of our ignorant selves, were pushed back upon the necessity of asking the land not so much what we could *make* it do, as what we *could* do, with what it wanted to do.

Today, twenty to twenty-five cows dot those thirty-odd acres of marginal hillside, land that has been our home for more than a quarter of a century. That land now provides virtually all the biological energy not only for our cows, but for our pigs, chickens, dogs and cats, our family, and our soil—with more to spare for the community at large. We learned by doing, which meant, often, that we learned by failing. Our goals weren't those of modern agriculture, and modern agriculture couldn't help us achieve them. Limited resources prevented us from trying to solve problems with money; harsh conditions forced us to fall back on those things damaged land can do on its own. It took a dairy cow to show us what these were.

Turned out to be a whole lot. Through this book, we hope to help you do the same thing with the land in your care.

CHAPTER 1

The Indispensable Family Cow

Twenty years ago, you could have driven back roads over every twist and turn of these hills, from one end of the county to the other, without scaring up a single hand-milked, grass-fed, homestead dairy cow. Once to be found on practically every small farm in the country, in the past seventy years the family dairy cow has nearly reached the point of extinction.

But that's changing. Take our neighborhood as an instance. If you pull up to Twin Cedars Farm around lunchtime on any day, you'll probably catch one of the children driving two or three Jersey/Limousin cows up to the barn for milking. At Stoneborn Pottery, you'd have to be there earlier; they milk their two Jersey-cross animals in the morning before breakfast. Just down Mt. Tabor Road, the folks at Holy Family Farm rise a bit later, but then you can find Brian and at least one of his five boys heading out on the side-by-side with a milk can and bucket. Down Knoxville way, Echo Farm milks Devons; Avemaria Farm, Shorthorns. On any of half a dozen other local homesteads the tale will play out with different details, but the same storyline. The family cow has made a comeback, and why? Because for much of the planet, the dairy cow is the pivot around which ecologically sound, small-scale food independence may be built.

Food First

The primary imperatives for human life may be reduced to just two: eat today; make provision to eat tomorrow. That is to say that, along with today's need for calories, ample and appropriate, we have also to make plans for tomorrow's meals. We need to farm, today, in ways that ensure our being able to farm tomorrow. That means producing food *and* fertility.

This is where the intensively grazed ruminant proves her value. Properly managed grazing ruminants on native pastures not only produce ample high-quality proteins and fats for the human diet in the forms of meat and milk, they also lay down an overall increase in soil organic matter and fertility to assure future harvests. And because good grazing produces denser pasture stands with deeper roots and shaded soil, holistic pastures will feature drought resistance, an active soil biome, and the healthy mycorrhizal relationships necessary for building complex plant nutrients. Not only ample meat, milk, and dairy products but also rich, bioactive soil fertility are the natural results of keeping a holistically grazed family cow, and more and more folks are finding this out firsthand.

Barriers to Entry

We knew none of this when, having bought that parcel of degraded land above the Ohio River, we began thinking about a dairy cow. We just wanted good milk for our growing family; but when it came to adding a dairy cow to our homestead, we got hung up on the issue of *practicability*. Just how much work were we talking about? And where was the time going to come from? What would this cow cost in money and forgone options? When all was said and done, how could we be sure this would be a practical choice? Like most families, we were already busy. We barely managed to show up clean and in nice clothes in all the places we were supposed to; more time at home in muck boots and dirty jeans seemed like retrograde motion. Obviously, for this cow thing to happen, we were going to have to reassess our way of valuing time and physical labor.

Modern Prejudices

Despite years of gardening and other homestead chores, we, like most people in industrialized societies, had an ambiguous relationship with protracted bodily work. Sure, "working out" has value in the general estimation, but the performance of physical *labor*—dirty, repetitive, physically demanding, productive work—makes modern people uneasy. We are so used to substituting mechanical labor for human effort in every possible situation that we end up wondering whether slow, sweaty progress is really progress at all.

But if the farm chores were to be done, then we needed to see time outside, time performing manual tasks, as of equal importance with income-generating work. If we were actually going to do what was necessary to make our farm fertile and productive, then the labor of producing milk and dairy products would have to be valued at least equally with whatever else we might have done in the same time.

Based on our experience with dairy goats, we figured that the care of a single dairy cow would probably take one person up to an hour, once or twice a day. While not a negligible chunk of daylight, it was time we could spare. Enjoyable as sports time, screen time, or hit-the-snooze-button time might be, we decided we were willing to forgo them in exchange for all the benefits we thought a dairy cow could offer us.

Personal Availability

Milking a cow meant committing to being home without fail at a certain time, every day, and this we thought we could undertake. After all, our lifestyle was already home-centered. Homestead chores come around each day like clockwork—not just milking chores, but all chores. An overindulged preference for ease, freedom, and unfettered spontaneity was going to be an obstacle to homestead bliss; so would a calendar full of off-farm commitments. Enjoyment of home, family time, and shared work is a must for the prospective dairyperson.

So we had priority reevaluations to make, lifestyle adjustments to consider, before we could find a place in our lives for a dairy cow. There would be other changes that would have to be made on the fly; we couldn't work out the details as a mental exercise. We had to commit to the cow, throw ourselves into the challenges, and then ride the waves—euphoric, when we considered the many gallons of milk suddenly added to our working resources; overwhelming, as we entered into a dramatic change of lifestyle for which we had no example—until we found our equilibrium. We had to commit to a family cow, before we could really know what that commitment would mean.

Access to Land

Land these days is wickedly expensive, so it's a good thing that in most climates just a small acreage will suffice to graze a dairy cow. Often, as few as a couple of acres will carry a cow clear through the growing season—for us, mid-April through early November—with enough standing grass to last for another month or more. And we're not talking about prime farmland. Grass is the default plant cover for about 40 percent of the land mass of the planet, including waste and abandoned acres. Land of little commercial value can often support a cow. Our own land is a case in point: with less than two acres of more-or-less level ground plus a few steep acres of poor pasture, we have fed our sizable family, including a milk cow, for going on three decades.

Financially, the family with a dairy cow may come from almost any income level; lack of funds need not be an obstacle. We see small cow-powered homesteads cropping up on cheap land, rented land, borrowed land, and even squatter's-rights situations. Many folks, like us, have established themselves on land considered "not suitable for agriculture." Some families in our neighborhood have been able to leverage better land, if it has minimal or nonexistent "improvements." The lack of development translates into lower mortgage payments, and the new owners, living simply in barn apartments, trailers, or unimproved older houses, are more easily able to move forward with their homestead dreams.

Money Well Spent

The cost of acquiring and keeping a cow, and the financial consequences of spending more time at home and less at paid work, far from being prohibitive, turn out to be incentivizing. You can buy a dairy cow or heifer for somewhere between a few hundred dollars to—let's admit it—the sky's the limit, but you can reasonably hope to add a good cow to your family for less than four thousand dollars, often much less. If she eats grass, her food costs nothing. Incidental expenses can be minimized. What will she do for you in return?

She'll mow your land and prevent it reverting back to weeds or woods. She'll convert your grass into milk, meat, and more cows, while composting and dropping 90 percent of what she eats back where she got it, thus adding organic matter and beneficial microorganisms to build soil and soil fertility. She'll douse the land in high-nitrogen fertigant (cow urine). She'll put gallons of milk on the table, and enrich your diet with pounds of butter, cheese, yogurt, and beef. Surplus milk will go a long way toward fattening a hog and increasing your chickens' rate of lay. And, if your grazing practices are holistic, each year your farm will be more fertile, more resilient, more productive, and healthier—and so will your family. Where would you find a better return on investment?

No Experience Necessary

One of the most exciting things we learned in our early years of homesteading was that the natural world doesn't require an instruction manual. This was a good thing, because, somehow, we had grown up around our grandfathers' farms without learning much of practical use. We began homesteading with only two assets: we weren't afraid of large animals, and we knew how to use a pressure canner. Of course, we had lots of experience carrying 50-pound (23 kg) sacks of commercial livestock feed, but we wanted to go in another direction.

Back in our urban homesteading days we had done a lot of research into the typical first two steps in homesteading: gardening and chickens. We weren't, at that time, concerned so much with organic methods and outcomes as we were with avoiding farming patterns that relied on purchased, off-farm inputs. We wanted to farm—which meant spending time on the farm, not at an off-farm job—which meant minimizing our cash needs—which indicated the low-budget farm goals. So we were disappointed and frustrated that every USDA bulletin directed us toward hatchery chicks, commercial chicken feed, and petroleum-based fertilizers, when we knew our grandfathers hadn't bought much of anything.

Where were the instructions for growing food without inputs? We couldn't find a single manual for doing things the way we wanted to.

Fortunately, despite the many skills and kinds of knowledge needed to practice natural homesteading, the aspiring homesteader can start out with remarkably little information. That's because the living world is governed by basic operating principles that can be observed by anyone willing to treat Nature with respect.

Much later, we were to find some hidden gems that would have been of great value to us in those beginning days. Sir Albert Howard, by many considered the father of organic farming, explains the links between soil, fertility, and human and environmental health. Joel Salatin encourages us to believe that basic grazing patterns, faithfully applied, will give the user all the education necessary for beginning a lifetime of grass management. Eliot Coleman elucidates the principles of natural soil fertility, four-season organic vegetable growing, and passive season extension, freeing our vegetable production from any reliance on off-farm inputs. Knowing that humans can use nature's own processes to enhance the natural productivity of a piece of land puts the budding homesteader on a trajectory for inputs-free success.

How It's Working

What we want is a bountiful, fertile, durable homestead that will produce food and soil from nothing but local sunlight and rainfall—and that's just what Nature wants to give us. Proper grazing of dairy animals on native forages not only generates an abundance of the highest-quality proteins, fats, and sugars, it also lays down a fund of organic matter and soil biota that, each year, increases the quality and quantity of those forages. When humans provide natural patterns of management, animals acquire greater familiarity with diverse forages, rediscover instincts for forage selection, and develop knowledge of therapeutic and medicinal forage species—they become "pasture wise." Each successive generation receives knowledge from those previous to it and adds discoveries of its own, reducing or eliminating the need for off-farm feeds, chemicals, or medications.

All this is free—with one condition. But it's a condition with a price tag. To receive all these benefits flowing naturally from the natural world, *we have to commit*. Because dairy cows need daily care. Grass management can't be done on autopilot. Portable fence systems mean walking, lifting, carrying—so, to use them, we have to use our bodies. Faithfully. We have to *abide* by our choices, and maintain them in the face of other, perhaps temporarily more attractive, options. We have to relinquish the option of having options; forgo the possibility that, when the mood strikes us, we might just not bother.

In exchange for this commitment we get, first of all, the best food in the world. The old saying "let your food be your medicine" finally makes sense in this context. It can't be described. Most people today have never tasted immediately fresh, minutely local, mineral-rich, nutrient-dense food like this. It's an experience that, alone, would be reason enough for the commitment.

But it goes beyond food. Managing cows on grass, daily, we acquire a deep and intimate knowledge of our place, our living nonhuman community, our land, our climate and soil. We serve Here; we become people of Here. We become native.

Natural beauty is another result of good grass management, and that beauty belongs to us. Not merely because we have title to the land; this is a beauty that moves into our minds, becomes our experience of the world, part of the fabric of our own being. Living becomes a joy and a privilege.

And aware, now, of the possibility of living as reverent parts of the many-lived Earth, we will not—we cannot—any longer be comfortable with behaving as autonomous units. We can no longer justify, or even imagine that we can justify, choices based on purely individual imperatives. We may no longer think that our personal rights can justly override other rights, not just those of other humans, but of the whole, living community that is Creation. We come to realize that the world is right only when people at least attempt to honor all rights. You can't make health by inflicting sickness; injustice imposed on another can never win justice for ourselves.

Grazing and milking cows show us daily that Creation is good, rational, and benevolent, but not *ours*, and not to be forced into shapes of our choice. The balance that, in Nature unassisted, involves predation, adverse climatic event, and cataclysmic boom and bust cycles, can, under human management, come to include appropriate animal harvest, weather resilience, and balanced stocking numbers. Humans, instead of being destructive parasites on the natural community, may assist in bringing new levels of order. But for this to happen, we must, as members of the natural world under our care, make ourselves subject to its laws. Then, having committed to the choice, we will find that, instead of losing options, we have gained the most important option of all: the option of staying here.

Community Benefits

Once we moved out of town, committed to our new lifestyle, and began to reap the harvest, we found that the benefits of our new life extended out in wide ripples to include our local community. Belonging to our natural surroundings, we found that we were also better connected to the human environment as well.

Today, when our low-income rural Appalachian county includes more small, grass- and dairy-powered homesteads than we can easily count, the bounty overflows from country to city in the form of herd shares, farmers markets, and community-supported agriculture. Workshops and evening lectures on gardening, meat production and butchery, and healthy food choices are a regular part of our social landscape; the winter lecture series at our village coffee shop draws folks from other states. Our Saturday market features produce from nearby small farms, as well as locally produced bread, beer, and cider. Farm vendors are a significant presence at our monthly street fairs; downtown, one business makes local, organic food available in a low-income neighborhood no longer serviced by even a single grocery store. These are all significant blessings—good food, pleasant social events, a healthy local economy—benefits flowing from our growing commitment to knowing and serving this piece of the planet.

And we're beginning to see a blessing of even more value: encountering this commitment to place, other folks, too, decide to stay and be local. Staying is a natural response, because this is such a good place to be. We stop waiting for the next job to take us somewhere else promising more money or prestige, or better entertainment possibilities. Our availability to one another stops being conditioned by the belief that the advantages of being somewhere else will someday outweigh the blessing of being together.

It turns out, in fact, that when we give up alternatives for the sake of choosing to milk grass-fed cows, that what we're actually choosing is a whole plethora of other good things. We choose to have the best, most nutritious, most delicious food—free, in great abundance. We choose the right to work, at least some of the time, with whom we want, when, and under what conditions. We choose to determine our own times for work, play, and rest, instead of clock-punching for someone else. We choose what we will grow, how much, and when we will harvest or slaughter. We are the arbiters of what our children will eat, and, to a significant extent, the kind of world our children will know, and how well they will know it. We get to build their inheritance.

It's a good life; it may even be that it is the only sort of life that can persist.

Slan Abhaile Farm

Damien and Melissa Murtagh,
Slan Abhaile Farm, central Virginia

"The day you have a ruminant producing milk on your land is the day you have immediate access to real, nutrient-dense food of all kinds," say Damien and Melissa Murtagh of Slan Abhaile (a name that means "safe home" in Damien's native Ireland). "Our dairy cow was the last of the livestock to be added to our homestead, but she has definitely been the most important—and beneficial. The only thing we regret about purchasing dairy cows is not doing it sooner."

The Murtaghs—Damien, Melissa, and their five children—started their homestead with, they recall, "zero previous experience," and "only resistance and judgment from our friends and family." Criticisms extended from the farm practices to the farm itself: the local abattoir assured them their animals couldn't be fattened without "at least six months on grain"; a "farm expert" told them their land "would not sustain much animal life without significant preparation and expense." Consulting agricultural journals confirmed that what the Murtaghs were trying to do was impossible.

Regardless, their dairy project has been successful from the beginning. Even though their land was "long neglected and overgrown," and covered with "dense stands of bush clover and stickweed," they began grazing an eight-year-old Jersey, with so much success that their herd quickly grew to include five cows. "Our only overhead is hay in the winter, and despite adding to the herd every year, our expenditure on hay has decreased. We track the starting date for feeding hay, and each year it has continued to go down."

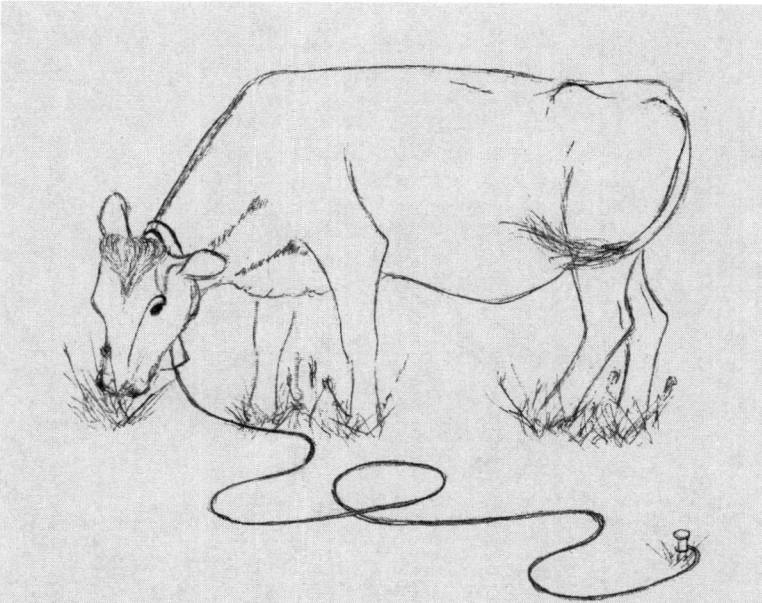

The Murtaghs farm solely to feed themselves. "Sending high-quality nutrition down the road didn't make sense," they say. "We decided early on that our focus was going to be regenerative food systems, and that means keeping nutrients on the farm as much as possible." As a source of food for the family, the dairy cows are an unqualified success. "Dairy is a staple in our house, and is present at every meal in some form or other," the Murtaghs observe, "and we buy almost no meat or dairy." With the integrated food security a dairy cow offers, they feel their homestead is truly a "safe home."

CHAPTER 2

Harnessing Sunlight

There's nothing like jumping in at the deep end. When we bought our first Jersey cow, Isabel, our own land was primarily trees-up-the-side-of-the-hill. Since there was no grass to speak of, we installed her in our neighbors' field. Steep, north-facing, and productive primarily of rocks and weeds, the most that could be said for it was that it wasn't woods. Isabel wandered at will, eating—well, eating whatever was there. In the winter, we fed hay. After her first calving, following standard instructions, we began offering her grain as well as hay.

Over the next few years, we engaged in a lot of observation. Experience was going to be our best instructor.

All Life Energy Is Solar Energy

Grass might be said to be the basis of agriculture, and human history is the history of grass management. Really. From hunter-gatherers, to pastoral tribes, to agrarian cultures, human survival over most of the planet for most of history has been firmly tied to the management of herbivores on grass.

There are good reasons for this. All life energy starts with sunlight, and that energy only becomes available to animals through the medium of plants. Forty percent of the land mass of the planet (and a much larger percentage of the habitable land) is some kind of grassland, which means that grass is the most prevalent source of solar energy capture available to us—by far.

Volunteer and Perennial

Now, when we say "grass," we mean weeds of all kinds, too, and pretty much anything else green that's not a bush or a tree. Look closely at an abandoned field and get an idea of how many species are growing out there—volunteer, native, or naturalized plants that were never planted, species that self-selected to be there. Among all this diversity, there are plants suited to virtually every climatic or impact situation likely to arise: flood or drought, heat or cold, there are species that love those conditions. Not only that, all this diversity means that solar energy capture is happening almost or actually every day of the year. It's going to happen wherever there is a leaf: from the tops of the tallest grasses down to the tiny forbs (nongrass herbaceous plants—think "weeds") almost at soil height, there are leaves of every shape, size, angle, and shade of green to gather in solar energy and convert it into food, all crowding together, trying not to miss a single photon. And these plants are maintaining their presence here—in fact, they're thriving—without cultivation, fertilization, irrigation, or any other kind of help from human beings. They *want* to be here—and that's the kind of energy, persistence, and determination we need to anchor our food systems.

In this business of harvesting food from plants, there's another serious consideration: *seasonality*. The fruiting parts of plants—those reproductive organs that feature so largely in the human diet, like nuts, berries, and fruits—are, with few exceptions, only available for part of the year. We can preserve them by refrigeration or processing, but a food culture heavily reliant on processing and refrigeration only shifts vulnerability from the food itself to food storage. It's clear there must be a loophole.

That loophole is grass. That's because when grass is harvested, it regrows—in many places two, three, five, or even more times per year. And grass can stand for long periods without losing its nutritional value, which means that, even in nongrowing seasons, it's food. That's a lot of nutritional energy, available all year long—an important advantage for people who require nutrients on a regular basis. Given the human preference for eating every day, all year long, a nonseasonal energy source is absolutely necessary—and that is what we get from grass.

Ruminants and the Conversion of Cellulose

It is in the context of grass harvest that the domestic ruminant—cow, sheep, or goat—is of irreplaceable value. Ruminants are characterized by their four-chambered stomachs, the first of which, the *rumen*, is

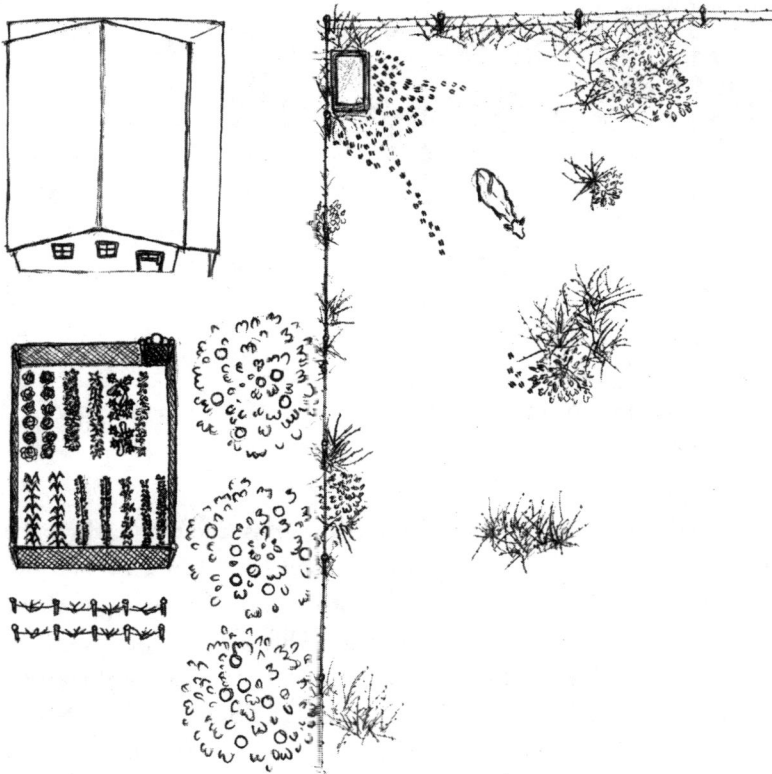

When grazing animals have unlimited access, forage is stunted and unfavored plants grow large and woody.

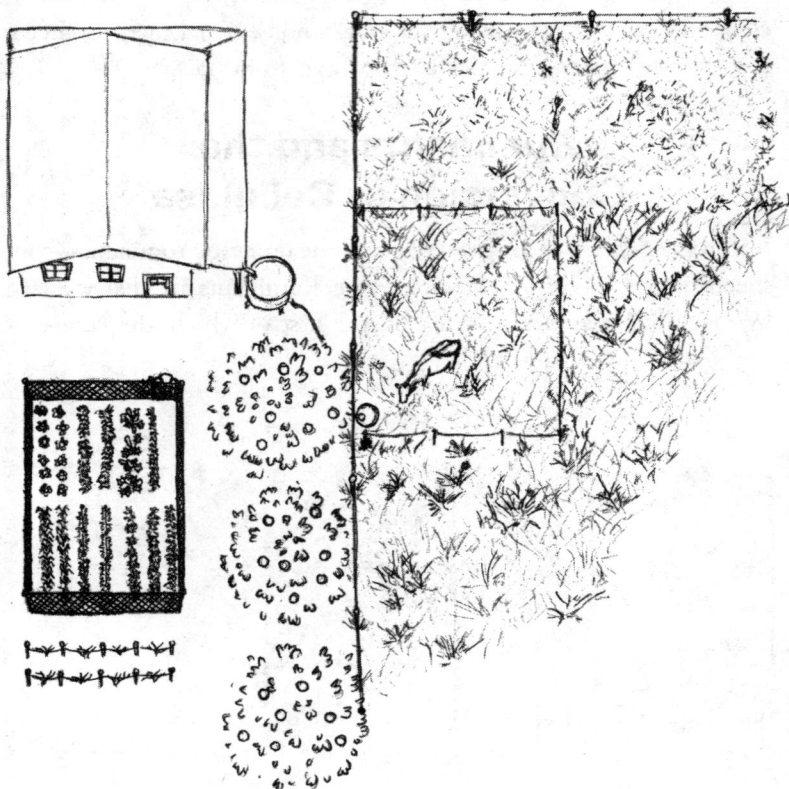

With small paddocks and frequent moves, forage can regrow
after grazing.

essentially a great big holding chamber for grass, and this is where the
first and most important step in grass conversion happens. Without
this step, unbelievably vast amounts of the biological energy on the
planet would remain unavailable to animal life—which includes us.

Remember, some 80 to 90 percent of all plant material consists
of cellulosic fiber (mostly cellulose, hemicellulose, and lignin)—the
tough, woody carbohydrate that makes up plant cell walls. Trees are
mostly cellulose, and so are the thousands of plant species that make
up grasslands. But, while cellulose is technically a carbohydrate—a
long chain of carbon and hydrogen, like sugar or starch—unlike
these simple hydrocarbon chains, cellulose is indigestible to humans.

Digestion of cellulose requires an enzyme, cellulase, and cellulase production belongs almost exclusively to things like bacteria, fungi, and protozoans.

Fortunately for us, ruminants have made a pact with these talented microbes. In fact, the rumen, first stop for everything the ruminant swallows, is a giant fermentation vat, and in it, those microbes feast on all the plant material the ruminant can swallow. They deconstruct it, releasing the energy in its chemical bonds, thus beginning the biodynamic cascade that provides life energy for so much of the planet.

And this is just the break we were looking for. Inaccessible by direct means, through the intermediary of ruminants the energy in fiber becomes available to the thousands of species that depend on cellulose conversion for survival—and that includes us.

Life Energy for the Future

Our food system is a two-pronged fork. It needs to feed us today, obviously; but, equally important, it needs to be preparing to feed us tomorrow. That means our food production can't deplete its sources; it has to restore the fertility taken out of the soil by what we harvest and eat. Think of fertility as a bucket of water; each day, as we dip some out for our present needs, we need to make sure something is refilling the bucket.

In nature, this is generally an automatic thing. When land produces food, that biomass is generally consumed more or less on-site, and the wastes and carcasses fall where they began, ready to re-enrich the soil. Humans are different, though; we rarely eat our meals (not to mention defecate, die, and decompose) on the site where our food was produced. Given that these days, each particular item of a meal may have traveled hundreds or thousands of miles to reach our plates, the energy and micronutrients that feed us have no chance of being returned to the soil they came from. So we eat today—but as we empty the bucket of soil nutrition, how can we make provision for refilling that bucket, so that we can eat tomorrow?

Let's go back to the grazing ruminant and her cellulose-digesting gut bacteria, and the whole community of living things that makes up a grassland. A grazing animal eats, defecates, and urinates all in the same place. Waste products are returned to their point of origin.

Moreover, because manure and urine foul grazed land, ruminants will avoid grazing again in that place for a period of time—which, when natural patterns are adhered to, allows the grazed forages to regrow. In other words, a second meal appears; because the land was manured and rested, Nature has a chance to refill the bucket. Doesn't this look a lot like solving our conundrum? *The same act that feeds the ruminant today makes provision to feed her tomorrow*—and that means it feeds us, too.

We hear a lot these days about how inefficient and even pernicious food animal production is, with its confined animal feeding operations and millions of acres of chemical-laden grains and soybeans, and those criticisms are more than just. Animals in feedlots eating grain are an ecological disaster. But ruminants self-harvesting the world's hugely abundant grasslands, and converting that fiber into proteins, fats, and sugars while simultaneously fertilizing those grasslands, belong to a different order of things entirely—you might call it blessings.

Food: fertility: more food. Win: win: win.

Cows Rule

Humans have been partnering with ruminants for millennia to build an abundant, resilient foodscape. Whether we're looking at the savannahs of Africa or South America, the steppes of Asia, or the Great Plains of North America, grasslands have played an irreplaceable role. Cows have not been our only partners; wild herds of deer, antelope, reindeer, buffalo, and so on have been and still are the basis of many foodways. Among domestic animals, sheep and goats outnumber bovines by a large margin. Each of these species can be the intermediary giving us access to the food energy in cellulose.

In different times and places, the human preference for domestic species has been a response to several factors. Our first step when we came to this farm was to buy dairy goats. We had several reasons for this choice, some better, it turns out, than others. The first reason was the best: we brought in goats because the land was growing goat food. That is, and always will be, the first consideration when choosing an appropriate ruminant for your homestead. Since your land is already growing a crop—native, perennial, volunteer, self-renewing, and ubiquitous—the next step is to bring in the animal that eats that crop.

Our second reason for choosing goats over, say, cows, was our belief that, for people getting started in dairying, goats were just going to be easier. There, we couldn't have been more mistaken.

In a contest between goats and cows for ease of husbandry, cows sweep the field. One difference becomes visible about five minutes after you bring the animal on the farm: cows will stay in a fence, and goats won't. Sure, we all know someone, somewhere, who says her goats aren't an escape issue, and such people deserve our congratulations, but most goats are Houdinis for whom an 8-foot (2.5 m) brick wall is scarcely an obstacle. Cows, on the other hand, can be confined with a single strand of electrified twine—and they're not too particular about whether it's electrified.

Cows, at least as presently managed by humans, are also much easier to keep healthy than are goats. There are a variety of reasons for this, mostly having to do with management styles, but the long and short of it is that you can manage a cow or herd of cows for years without using meds or wormers, while by far the majority of goat keepers have regular recourse to a whole pharmacopeia of chemicals to keep their animals healthy. Further, unlike goats, with their multiple-offspring births and the tangles of legs and heads that can ensue, a well-managed cow almost never needs help at birthing time. Given all this, we can't help feeling the only thing that's really easier about goats is lifting them off the ground.

There are other reasons we're cow people. Taste is one—we just like cow's milk better than goat's milk, at least for drinking (chevre is another matter). The abundant, self-rising butterfat in cows' milk is another: cream without machine separation. But there are two more reasons that would, even if we left our food preferences aside, incline us to keep cows whenever the native plant communities made it an option.

The first is in the nature of forages. If you're growing briars and shrubs, you're growing goat food, and you should keep goats. If you're growing grass and forbs, that's cow (or sheep) food. But over time you'll notice an important difference in how these forages respond to impact (grazing). Grass likes to be grazed, and gets thicker and lusher under good grazing practices. Bushes and briars, by and large, will be damaged or eradicated by regular grazing. The more they are grazed, the less there will be; it's like planned obsolescence. There is more than one reason for this difference in plant response, and that discussion belongs in another book (one we're presently working on), but it is a fact that good grazing of grass pasture is going to produce more forage over time, and browsing brush mostly won't.

The second reason we lean heavily toward cows is a simple preference for volume. We want a lot of milk, and we want it all year. Milk—energy derived from completely local sunlight—is so valuable on the farm that it's hard to imagine having too much of it. A single cow may easily provide multiple gallons of milk per day on average throughout the whole year or even several years, a feat that would take more than a few goats, which not only produce less milk but have a shorter lactation period. We don't mean to say it couldn't be done, and if your land produces more goat food than cow food we'd encourage you to do it, but these are some of our own reasons for preferring cows—and why the rest of this book is going to focus on milk production from bovines.

Grain: Yes or No?

The great gift of ruminants, as we have seen, is their ability to access and convert the solar energy trapped in the chemical bonds of that superabundant organic molecule, cellulose. This gift makes them superstars in the history of human alimentation, and there is no overestimating its value. Their food is perennial, volunteer, and ubiquitous. Grain, on the other hand, is an annual that requires—at least in our present modes of production—enormous inputs in the forms of labor and petrochemicals. Moreover, *feeding grain to ruminants is just not necessary*. You can offer grain to a ruminant, if you want it to get fat faster or produce more milk for longer, yes, but ruminants don't have a natural, metabolic, dietary need for those concentrated starches. And feeding grain to an animal that Nature has designed for cellulose digestion is not only wasteful, it can bring with it a whole host of health issues—not to mention being expensive.

But know this: If you elect to feed your cow only grass—and it is to such people that this book is primarily addressed—you're probably going to get some pushback from other cow people, because in the twenty-first century, feeding ruminants grain is deeply embedded in our agriculture, and people won't fail to tell you about it. Your veterinarian, your neighbor, the guy at the feed store, and any stray passerby who happens to feel talkative is going to wonder why you're "doing it wrong." They'll be able to tell you how much more milk you'd be getting, or how much faster your steer would be growing, if you would only feed grain. Well, let them talk—and then go on keeping your cows on grass, knowing that this is what they're designed for, and the best outcomes will only follow, indeed *can* only follow, when we honor natural designs.

Milk Feeds Everything

Now, let's talk about milk, and the qualities that make it indispensable.

Abundance. Remember, milk comes from cellulose—that primary and superabundant source of food energy for the planet—and ruminants will give us milk every day.

Nutrition. Milk is a whole food—every mammal on the planet begins its life on a diet of nothing else. Everything a baby mammal needs is present in this one perfect food. Not only does milk provide all our macronutrients, like proteins, fats, and sugars; grazing animals also pass on to us the nutraceutical phytochemicals (read: plant substances that are good for your health) from their forages. That's as though you, yourself, were eating a diet including hundreds of species of plants—some serious diversity, with all the attendant benefits.

Versatility. Milk feeds everything. Really. Calves, obviously. People, and not just as drinking milk, but as all the probiotic milk ferments, like yogurt, kefir, quark, butter, and the thousands of kinds of cheeses. And then there's the homestead. Let's think about it.

Human beings have been milking ruminants for thousands of years, their lives deeply enriched by access to grass energy in the form of milk—lots of milk. More than people can use while it's fresh. So, what happens to the gallons and gallons of secondary products (skim milk, buttermilk, whey) when milk is processed for storage? What do we do with all those wonderful, perishable nutrients? We convert them into other forms, by feeding them to pigs, who turn them into long-term, on-the-hoof, delicious protein storage; and chickens, who use it to make eggs; and predator- and pest-control animals—dogs and cats—so they know where home is but are still hungry enough to hunt. Turns out milk is the high-nutrient supplement that ensures farm-fed animals get the proteins and fats they need for health and growth, and daily, or even twice daily, our cup, and theirs, literally overfloweth.

Homesteading—the word means "home-stay" or "home-stand"—is about staying put. We arrive at a good place and want to settle down, and grazing animals is how we do it. All hail to the kitchen garden, bravo to the orchard, warm kudos to the vineyard—these adaptations on the local terroir are greatly to be blessed. But the strength, the durability, and the character of the land are in what the land is already doing, naturally and without assistance—and that's growing grass. Grass makes soil; grass holds soil; grass produces soil fertility; grass protects soil. And ruminants turn it into milk, daily, forever. Milk is the homestead's lifeblood. And it's yours for the harvesting.

From Wasteland to Home Land

Kaleb and Christa Hanshaw,
Highwall Farm, Logan, West Virginia

Kaleb and Christa Hanshaw and their children are leading lights in the Appalachian strip-mining reclamation program Coalfield Development, which has made enormous strides in regenerating severely damaged soils.

"Our program began with the regeneration of damaged and strip-mined land using pastured pigs and poultry," Kaleb says. Two years ago, they added all-grass, holistically grazed dairy cows to the mix. "The results have been nothing short of remarkable, with noticeable improvements in soil health, pasture quality, and ecosystem balance," he adds.

Milk has provided a new source of nutrients for the whole farm. "The introduction of dairy cows has transformed the way we think about farm sustainability," the Hanshaws share. "One of the largest challenges in modern farming is the cost of and reliance on outsourced animal feed. By integrating dairy cows, we've reduced that dependency significantly. The abundant milk serves as a cornerstone for the entire operation; not only does it nourish our family, it also supports the health and well-being of all the animals on our farm—pigs, chickens, and even our dogs—a shift that has improved their overall vitality and reduced their need for commercial feed."

In response to these successes, the Hanshaws have increased their herd size to six dairy cows with calves, and now consider their animals indispensable allies. "Each cow has her own personality, and it's impossible not to grow attached. Whether it's feeling the rhythm of their breathing while milking or learning to anticipate their quirks, these animals become so much more than livestock—they become partners in the work of regeneration."

Kaleb believes the cows are there for the long haul. "This way of farming is more than just a method; it's a philosophy. It's about creating abundance, not just for our family, but for the entire ecosystem we steward. Dairy cows have taught us how to be better farmers, better teachers, and better stewards of the land. Their impact extends beyond the milk they provide or the nutrients they cycle through the soil; they are at the heart of our farm's story."

CHAPTER 3

Land and Experience

We hope we've been painting you a picture that's pretty irresistible. Grass-plus-cows-equals-food—the best, in copious plenty—is a pretty good story. Lots of people are going to want to write themselves into the script, and that's just what we would encourage them to do—and then stay in the game, despite the challenging times ahead. We're all in the same boat: our culture doesn't teach us how to do this, and it won't be in the bleachers cheering for us, but as budding farmers we need to stand firm in the knowledge that good grass management has always worked—and it always will.

This was a lesson we had to learn the hard way. As it happens, both of us grew up around beef cattle. Our dads, farm boys from the Depression era, began life on Southwestern dry farms where their parents raised large families on just what the land would produce. Although our fathers ended up in city jobs, they kept one foot in the country, so we grew up around beef cattle: herding them, putting up hay for them, feeding and doctoring them, and (of course) eating them.

But we *didn't* grow up milking them, nor did we take an active part in the planning around those herds. As far as agronomy went, we were as ignorant as any city kid. When it came to being responsible for the well-being of our animals, making decisions about their feeding and health, we were still going to endure a good many sleepless nights. So we sympathize with the doubt—and self-doubt—that may shackle the erstwhile homesteader. None of us knows this playbook.

How Do I Dare?

Adding a dairy cow to the household is certainly a break with the standard way of doing things today, a rejection of an industrial, digital culture dedicated to conformity. And we're used to conformity: Americans dress, eat, recreate, exercise, and think pretty much according to code from the time we're clad in our first disposable diaper to the moment we graduate from high school or college or graduate school and sign on with our first employer. Following instructions has worked itself into our epigenetics, so it's small wonder if our sense of safety and self-preservation is correlated to our adherence to code.

In the face of this, how can anyone even think of taking on a family dairy cow? Chances are good we have no land, no livestock, no education for this undertaking, and no experience. Most of us won't have much in the way of discretionary money; almost all of us have debt. If we think we know what a farm looks like, we're probably wrong; if we think we know what kind of equipment and infrastructure we are going to need, we're wrong for sure. Keeping a dairy cow is a major departure from the conventional road to success, the conformity we've always associated with security. It's a step for which we have no prior experience—so it's reasonable to ask, to begin with, if it can even be done.

Land

Land is the first, biggest, and most expensive obstacle to embarking on a family-cow career, and it's the one with the most misconceptions around it. A month seldom goes by in which we are not consulted by least one person who says her most insurmountable obstacle to beginning a much-desired dairy future is lack of land access. "I don't even know where to begin," is the frequent cry, and, maybe even more often, "I can't find a farm anywhere."

Well, maybe the first problem is that this person is looking for a *farm*—or thinks she is. What does a farm look like—especially a small, subsistence farm anchored to the land by dairy animals? Whatever picture she may have in her head, it's almost certainly wrong. Let's take apart the usual assumptions one at a time, so we can dismiss them all.

What Is a Farm?

A farm is simply land under human management for the purpose of growing some things for human use. It's not a set of buildings, or a network of fences, and it's certainly not a volume of open, cleared, tilled, or mown land, however much these things may look, to the inexperienced, like a farm. For your farm, you are going to need a piece of land that gets sunshine and rainfall; there must be access to drinking water for livestock; and it can't be so steep that there is nowhere an animal can stand upright. Within limits, almost anything that meets these criteria is a farm-in-potential, regardless of whether the piece of land is strictly speaking "rural" or is the vacant field behind a loosely zoned suburban lot.

The less desirable the land is for other uses (development, agribusiness), the less competition there will be for it—and the lower its price. And, as it happens, such little pieces of unwanted land often have more going for them than meets the eye. The same qualities that have left the land unused for commercial or residential purposes may increase the possibilities it offers the homesteader who wants to keep a dairy cow. A marshy area may provide midsummer grazing; an east-facing slope offers shelter from prevailing storms; a variable topography offers multiple micro-ecosystems to the future gardener and orchardist. Who says farms must be cleared and flat?

There are, of course, *some* basic requirements for a piece of land to be useful as a homestead. Water access of some kind is necessary, but this need not be pressurized county or well water. A reliable stream can often be utilized for stock water in ways that don't create negative ecological impact; or captured roof water may provide what is needed. Some grazing is possible in wooded areas, but for serious grass harvest you won't want to be confined entirely to forested land. Contrary to rumor, there's nothing inherently wrong with cutting trees. Much land that is now wooded was, until recently, cleared for pasture or cropping. Restoring some of that area to use can be a great idea, so long as you avoid doing it in a place where you risk destabilizing a hillside or creek bank. You also need to have access to the land, of course, and it's nice if you're not carrying every bucket of milk and bale of hay by hand over a deep gorge on a narrow footbridge—you're going to be doing this once or twice a day, so it's important to keep expectations realistic.

Do I Have to Own the Land?

The homestead dairy is within reach even if you can't find an affordable piece of land for sale; land ownership is overrated. We don't need title to the land, we just need the use of it, which in these days of heinously inflated land prices is good, good news. Forty years ago we ourselves had an unaccountable prejudice against using someone else's land—"What! Improve land we don't own? Not on your life!"—but, fortunately, we got over it, and for the past twenty years we've had an informal land-use agreement that gives us access to thirty acres of pasture for which someone else pays taxes. Our arrangement is cordial and has been the foundation of a deep, lasting community relationship. And we've had the use of that land—to graze on, to harvest from, and to learn with—in exchange for good stewardship and some shared beef and dairy products. Where's the downside?

Honestly, we'd keep a cow if we had no land at all. Urban dairy cows have a long and time-honored history, speaking to just how valuable the grass-to-milk/meat/manure transformation really is. For the cost of a light bale of so-so first cutting hay you get multiple gallons of milk per day, plus enough manure to make the best garden in the city. So, no, my friends, you don't have to own a piece of land to make it worthwhile keeping a dairy cow.

How Much Land Do I Need?

Just how much land are we talking about? A cow's requirements in acreage varies with the forages and climate. Grass growth and regrowth are dependent on the volume and frequency of precipitation each year. "Brittle" environments, where rainfall is very seasonal and long, dry periods are common, require more acres per animal than do regions with more forgiving climates. For example, here in northern Appalachia, a Dexter–Jersey cross can graze pretty much all year long—and, yes, we do mean winter, too—on a couple-three not particularly productive acres. In the Shenandoah Valley, Polyface Farm runs cows at about twice that density. A farm in Tehachapi, California, where the rainfall is very seasonal, needs a much higher acres-to-animals ratio—something closer to ten to one. And in the mountains above Gallup, New Mexico, homesteaders are using their dairy cows to build pasture grass communities where presently there is only rock and gravel, feeding baled hay on the ground and leaving plenty of waste hay and manure—so, for the moment, their acres are limited by their stock numbers, not the other way around. Ultimately, keeping a dairy cow is less about how much land you can access, and more about how cleverly you use it.

On the other hand, there *is* such a thing as too much land. Why? Because good management requires attention, and if your attention— and your animals' impact—gets spread too far, the land suffers. If you buy a hundred acres in the Cumberlands to keep two dairy cows and a steer, you've got more land than you—or they—can even begin to handle. Either you'll try to manage all of it, and spread your cows' impact too thin, or you'll limit your management to the acreage your cows can actually impact, and the rest of your land will suffer from neglect. Bigger is not necessarily better for the subsistence farmer; smaller acreages offer plenty of potential.

Grass Management for the Completely Ignorant

The next thing is to know what to do with your land. Leveraging sunlight via grass through the guts of a cow is the original, God-given, golden-egg-laying goose, but there is a catch: *appropriate* grass management practices are a must. Good grass management means holistic, rotational grazing, cows moving from paddock to paddock on a daily or even twice-daily basis, with regular observation and adjustment. Unfortunately, we learned none of this stuff in school. Our fathers' herds grazed on large pastures for whole seasons at a time. So when we set out to eliminate grain feeding from our own homestead *without* putting damaging pressure on our pasture, we needed a crash course in holistic grazing.

Or so we thought. Actually, what we really needed was some good advice, and we found that where you usually find most good information, in books and conversation with experienced people. Joel Salatin's *Salad Bar Beef* gave us the basic principles—"hold 'em tight and move 'em fast" (calling for short duration, heavy impact grazing)—and encouraged us to believe the results would be worth the trouble. Allan Savory's multi-million-view TED talk on environmental degradation and regenerative grazing was breathtaking confirmation. And a wonderful local coalition of Appalachian farmers, the Eastern Ohio Grazing Council, gave us our first look at some good grazing happening in our own neighborhood. Online farm suppliers ship fast—because they know that by the time a farmer places an order, she needs it yesterday—so it was literally within a few days of our first exposure to holistic grazing that we were out in the pasture, setting up little white-string paddocks for our cow, Isabel.

Like riding a bicycle, good grass management is something you have to learn by doing. There are a few fundamental rules to keep you on track for consistent pasture improvement, and some very flexible guidelines to help you get started, but all the real learning happens in the field. Who teaches you? The cows teach you, the grass teaches you, the weather and terrain and seasons teach you, and most of all, observation and reason will teach you most of what you need to know about cow nutrition and the million fascinating things there are to know about grazing.

Grass management is not a discipline that offers short cuts. While you can make a perfectly adequate start with just a handful of precepts, avoid anything that looks like a slick trick. Working with nature is forging relationships, and relationships only grow with presence, time, and attention. Gadgets that replace human presence—and there are some very clever gadgets out there—are mostly obstacles to the actual hands-on, boots-on-the-ground experience that is how you build a relationship with your land and the living things on it. The time it takes to learn grazing isn't a disadvantage, it's one of the blessings your homestead and your cow are going to give you—but to receive those blessings, *you have to be there*.

The Time Factor

For the first twenty-odd years of our marriage, babies arrived with almost clockwork regularity; about the time the current baby turned two, Mom received her first intimation—chronic nausea—that another little Dougherty was on the way. We did that eight times. We home-birthed; we homeschooled; we homesteaded. We cooked from scratch. Every spare second had three things we could have put in it. Obviously, we needed a dairy cow.

Maybe your life is like that: packed with commitments, wall-to-wall with obligations and unavoidable time allocations. So how is it remotely possible to find room for a substantial chore that comes around every single day without fail? Thinking of getting a milk cow is kind of like considering a new workout routine; it's fun to imagine we'll start swimming a mile every morning before breakfast, but that's about where the fun stops. There's never really going to be a good time to adopt a new self-torture regime. If you're going to take on the chore of milking and pasturing a cow every day, maybe even twice a day, the first thing you'll want to know is, what's in it for you?

The answer is about more than just dairy products. Sure, there's some milk happening in there somewhere, but whenever you let an animal engage in natural behaviors like walking around, harvesting its own food, and reproducing, there's a lot more going on than you might imagine, and, in this case, it's pretty much all great.

What's in It for Me?

Wherever grass grows, ruminants are the natural providers of myriad ecological services.

Lawn Care

If you are thinking of getting a dairy cow, you probably have a little land you want to put her on. What's going on there right now? If you're not mowing it occasionally, then it's getting overgrown and needs some attention; if you are mowing it, then it's getting the attention it needs from your personal labor. Enter the dairy cow. Part of the work you will do for her is to move a piece of white string (her polytwine fence) a few feet forward every day to give her a new paddock. Sure, you just added a chore to your day, but what's she giving you in return? Not just mowing services, replacing your hours on a tractor, but also fertilization and irrigation services. A cow can produce over 100 pounds (45 kg) of manure and urine every day, and your holistically grazed cow will even spread it where it will do the most good.

Weed Suppression

Holistic grazing patterns have long-term weed-suppression effects as well. The plants that are of the most value to the grazing animal—and, not coincidentally, these are also the ones that respond well to the graze-and-rest cycles of holistic management—flourish as a result of your daily visit, crowding out less palatable, less nutritious species. Moving a fence does more to improve pasture composition in the long run than the most state-of-the-art soil amendments, seed cocktails, and no-till drills.

Irrigation and Flood Control

Improved ground coverage means rainfall events soak in instead of running off. Hydrated soil remains bioactive, increasing the volume and food value of forages grown, and resisting dry-season dormancy. If there's surface water somewhere in that pasture—a pond or stream—periodic grazing can be the means of correcting erosion problems on the banks. Yes, standard wisdom is to keep animals away from riparian areas, but under good holistic management banks are graded and grassed over, protecting and improving waterways.

Ecosystem Enrichment

Adding a new animal to the ecological community will attract or increase the presence of other species, building interest and resilience into the biome as worm populations soar and insect-eating birds increase, creating niches for new predator species, and so on—all for the trouble of your daily fence-moving chore. Don't underestimate the value of biodiversity, either; it's not primarily about tourism, it's about resilience and abundance.

Food

And then, of course, there's the milk. Did you forget the milk? By adding a dairy cow to your daily routine, you'll be adding hundreds of gallons of milk a year to the available fuel on your farm. People food—like drinking milk, butter, yogurt, kefir, labneh, sour cream, cream cheese, mozzarella, and aged cheeses—and also an abundance of food for your pets and livestock. Milk, especially the by-products of dairying like skim milk, buttermilk, and whey, is the secret to how our grandparents all the way back to Noah fed their pigs and chickens. In offering them surplus dairy, Grandma knew she was raising the healthiest pigs—which, in turn, became the healthiest pork. Chickens turn extra calcium and protein into eggs and meat. Dogs and cats still have appetites for hunting when dinner is a bowl of milk in the barn. How much money will you be saving with all the milk your cow is giving you? Considering that conventional grocery supplies are delivered from distant sources on a just-in-time basis, how much food security does she represent? And this is just some of the payback for keeping a family dairy cow.

Buy-In

It's obvious that a homestead dairy cow has a whole lot to offer. There's no getting around it, though: she's going to demand some labor and faithfulness in return, and you want to be sure before you undertake it that you're doing the best thing for your family—and your cow. This is a matter for the family in full council, because in the days going forward, you'll need to have everybody on board. A dairy cow is something you don't want to do completely by yourself, and it's not going to work if your family is dragged into it kicking and screaming.

There are practical questions that need to be addressed. Where in your day do you imagine yourself fitting this new chore in? It's certainly possible to switch up your milking times seasonally, but from day to day you need—and your cow needs—an established milking schedule. Milking and moving fence, and then processing the milk, are going to take at least thirty minutes most days, probably longer, so you'll have to plan for it.

Who will be milking? You need a minimum of one regular milker and one backup milker, because the most devoted dairyperson in the world still gets sick or wants a break now and then; and, of course, life happens. Children can make excellent milkers—one mom we know shares the chore with her nine-year-old—because the primary requirements are strength and endurance; but this is not a chore to turn over entirely to a child, especially not in the beginning. Keeping a dairy cow is a huge blessing, but it's also serious business, with your milk quality and your cow's health dependent to no negligible degree on the observation and experience of the person who is going to be there every day, making small but significant decisions. That responsibility should rest with someone of mature judgment.

When Do I Get a Break?

This is the number-one FAQ, and the major stumbling block for anyone even remotely considering a family dairy cow. Fail to answer this question, and the subject is automatically closed. Relax, though, because you *do* get a break; you're in charge of the breaks, and you can make this as easy or as rigorous as you like—so long as you accept the trade-offs.

First, there are some milking hiatuses that are typically built in. Over the course of a lactation, twice-a-day milking goes down to once a day, as the needs of the nursing calf (if there is one: see chapter 10, Calves and Calving) increase, or as your cow's production decreases. And most cows are "dried off" for eight or more weeks before a calving, so there can be two months of no milking built in.

Further, there are breaks you can choose to install, even spontaneously, if you like. Herein lie the trade-offs. The various patterns of calf sharing (milking while letting a cow raise her own calf) can give you anything from a once-a-day schedule to breaks lasting whole days or weeks. They're all great ways to keep a milk cow, to raise a good calf, and to maintain a schedule built around your family timing; just know that they'll probably reduce your cow's production to a certain extent, as well.

Give serious thought to your family's present commitments and conditions. Even with family buy-in, are the demands of a dairy cow reasonable for your family right now? We know more than one family in which the primary milker is a mom with toddlers and infant in tow, and our hats are off to them all; but this is only possible because those families are already unconventional in their allotments of time and energy. Often they also homeschool, breastfeed babies, and do most of their socializing on-farm; usually, Dad is good in a pinch for a share in the milking chore. For families with long-distance school or work commutes, time-consuming sports commitments, and sundry extracurriculars, adding a substantial chore to either spouse's daily list may not be reasonable.

Finding Help

Your dairy cow is going to be your new best friend, but if you're a single person living alone, it will be good if you have another best friend, or at least a vitally interested and committed neighbor, to invest time with you in this life-changing development. We're seeing more and more multihousehold cow cooperatives, and while we were once skeptical of this arrangement—on the grounds that the world only holds just so many people crazy enough to commit to dairying and what are the chances they'll be our neighbors—today we are ardent fans of the shared dairy cow. Yes, care must be taken to see that everyone in the cooperative has the necessary skills and understands their role, but with the benefits as great as they are, we expect to see this arrangement become more and more common.

It's a given that few people in this day and age will be bringing extensive prior experience to their new dairy endeavor. Most will never have squeezed a cow at all, even if they know at which end to start. Milking is a skill that must be learned by doing, and while proficiency comes pretty quickly, speed, by contrast, is a matter of long practice and muscle memory. There's no denying you'll enjoy more peace of mind if you can have an experienced eye, hand, and voice close by while you're in the learning stages. Milking mentors may not stand on street corners holding signs, but if you can find someone in your neighborhood who has experience with all-grass, holistically managed, hand-milked cows, cultivate a relationship with that person.

Conventional dairy folks—those whose experience is all or primarily of commercially managed, grain-fed, machine-milked animals—also have a great deal of experience and information, but be aware that their methods and goals, radically different from yours, often lead to advice and conclusions that, while eminently applicable in their own sphere, are counterproductive in yours. Vets and commercial dairypeople are unlikely to understand your preference for less milk in a more seasonally nuanced supply, based on grass, over hyped-up performance based on confinement and grain feeding. The problems of grain-fed cows are different from grass-fed ones; the appropriate conformation and norms of health will also be different. Advice and help are great things; just make sure they apply to your farming model, or they'll be the reverse of helpful.

Thoughtful Life Adjustments

With your family, friends, or neighbors fully on board, you've nailed down the human factor. There's also the issue of material preparedness.

If you have grass, you can feed a cow; that's the first and most important requirement. A certain minimum of infrastructure and equipment needs to be in place, hopefully before you bring your cow home; in chapter 5, Settling Your Cow In, and chapter 6, Milking, we go into the things you'll need to make both of you comfortable. And you'll need to have given at least a little thought to a question that you'll soon be asking yourself every day: What am I going to do with all this milk? This is like asking a child with a full Easter basket, "What are you going to do with all that candy?"—a delightful conundrum, a blessed dilemma. Still, it's a question you'll want some answers to from the very beginning, and chapter 7, Milk Handling, looks at milk processing, storage, and destinations.

Adding a dairy animal to your farmstead is not a decision to be made on impulse. She is going to change your life radically, and, in the beginning especially, it will help if you can remind yourself every day of all the good reasoning and planning that went into this project. You're sure to have second thoughts; it will help a lot if you start out with plenty of first thoughts.

But, once you've embarked on your dairy life, the rewards are enormous: to your land, to your food quality and stability, certainly, but also to the quality of your daily living. Dairy cows are big, gentle, companionable animals. When you work with them daily, they become partners, personal others. If you have a beloved pet, you already appreciate the depth that an animal–human relationship can have and the richness it adds to life; your collaboration with a dairy cow adds another, powerful dimension to that experience. Working with a nonhuman for your mutual well-being goes a long way toward reforging your connection, your belonging, to Creation. We can't help concluding that this kind of collaboration fulfills a deep need—psychological, physical, and spiritual—that is neglected in our modern culture.

Personal Changes

Adding a holistically grazed dairy cow to the land under your care is the single most productive, regenerative step you can take, but it doesn't come without a price tag. Without attentive care, your land and your cow can be harmed—and it will be your fault. This is a moment when self-knowledge is indispensable. Make a good examination of conscience: Are you going to commit to this step? Will you make the well-being of the farm, the family, and the cow a real priority—every day, at milking time? Are you in a place where it's reasonable to think you can add a new commitment to your life, and then keep faith with it? If you are already overwhelmed, maybe this isn't good timing. Care, attention, and, dare we say it, love, are requisites if you are going to earn the rewards—fruitfulness, fertility, belonging—of keeping a dairy cow.

When you have made the leap, though, the blessings become apparent. One, unexpectedly, is that this new chore actually improves the organization and productivity of your daily schedule. With several small children, one off-farm job, and a long list of domestic and farm responsibilities, we found that having a nonnegotiable commitment every day gave us something to build our schedule around. One person rising early for the morning milking and coming in hungry for breakfast meant there was a certain gentle pressure on the rest of the family to participate. We found ourselves with an established time for rising; jobs were delegated, breakfast prepared promptly. The meal together gave us time for planning of shared work, and our days began to have a unity and productivity we'd never experienced, but for which, we realized, we'd always unconsciously longed. We began behaving like a *family*.

And as a family we found that dairying—feeding ourselves from this place and from our care for it—was giving us something else for which we had no previous experience, but which seemed to reach some inner need: belonging. We were building connections to *this* land. Its woods and open fields, its climate, the plants and animals that lived here, and the other humans settled in this place were revealing themselves as an unshakeable reality that put solid ground under our metaphorical as well as our physical feet. We were becoming native—and it felt like coming home.

Keeping the "Family" in the Family Farm

Brian and Johanna Burke,
Holy Family Farm, Richmond, Ohio

What do a doctor and his family do when they are really passionate about clean food and natural fertility? They buy a dairy cow. Brian and Johanna Burke homestead Holy Family Farm, a patchwork of six owned acres plus five acres borrowed from two different neighbors, where they raise their six children, three cows, three dozen chickens, and seasonal livestock of all kinds. Their gardens supply them with food and medicinal herbs. Nutrition is key to healing, and Brian frequently recommends raw milk and organic whole foods to his patients, as well as the benefits of an active homestead life.

The Burkes' children are intimately involved in every aspect of farm life. Milking time finds Brian and one or more of the older siblings headed down to the barn carrying buckets and milking cans. Moving temporary fencing for holistically grazed paddocks is a daily chore often entrusted to one of the children, whose pasture knowledge increases in tandem with their parents'. Even the very young children have age- and skill-level-appropriate chores for which they are responsible, and chore time finds them all donning muck boots and heading out the door.

Certain the homestead setting is a great place for raising healthy, independent children, the Burkes actively assist other families getting started in homesteading. Through hands-on workshops and farm tours, they help other families see the practicability of family-scale dairying. Keeping investment at a minimum is especially important.

"One of the best things we did in the first few years was to keep as much as possible of our infrastructure temporary," the Burkes recall. For instance, a twenty-year-old barn came with their homestead. "The inside was a complete blank," says Johanna. "Instead of building permanent stalls or stanchions, we built in ways that would be sturdy, but could be changed without too much effort if our needs were different in the future. This strategy has worked remarkably well, as our needs for that space change frequently throughout the year."

CHAPTER 4

Acquiring a Cow

Flashback to October, 1998, and the day we bought our first dairy cow. Isabel was eight months old and small enough for Shawn and the boys to lift into the bed of the pickup; when she got fidgety the boys opened the back window, drew her head inside, and covered her eyes with a jacket. It was Hallowe'en, and a state trooper shadowed us halfway home, wondering, probably, what kind of prank we were up to. This was a watershed moment: Isabel would be with us for twelve years, and, with her blessings, limitations, and downright faults, would give us a thorough, if sometimes stressful, education.

How does one go about finding a dairy cow? Shopping is in the American genome, a homozygous dominant gene reinforced by the epigenetics of use in this age when our every need must be translated into money. We are expert consumers, whether the purchase is a jar of tallow balm—let's see, F-Balm or Toups & Co.?—or a used pickup— what's the Blue Book on a late F-150? Still, one item most of us have never shopped for is a cow, and there are a few things you'll want to know before you start.

What Is a Dairy Cow?

First, what makes a dairy cow, a dairy cow?

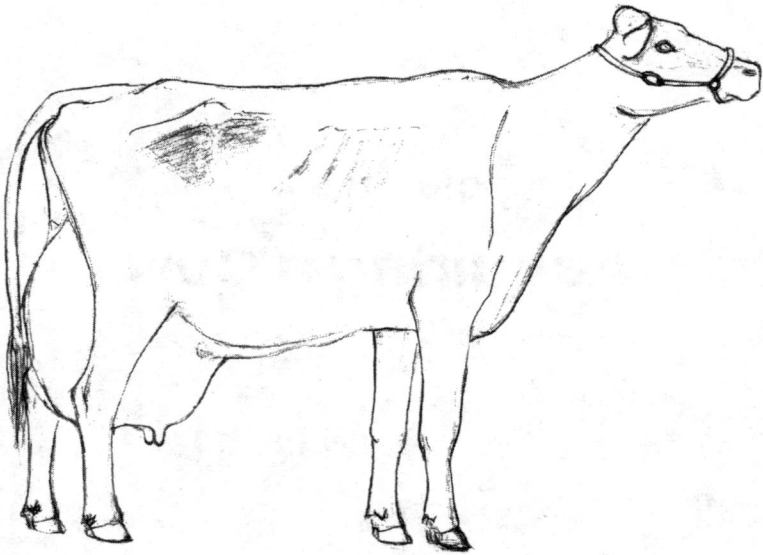

A commercial cow, with her massive udder and bony frame, is designed for grain conversion.

Our ancestors were practical people. In the process of domesticating the primitive *aurochs*, progenitor of the modern European cow, they selected their brood stock according to several desirable (to humans) traits. First, of course, was a sound and enduring constitution (read: health) without which none of the other traits mattered. Then came the three functions humans desire in a cow: a beefy carcass; the ability to pull implements of cultivation; and milk production. For most of history, cows were bred to be useful in all these capacities, and breeding just for beef or milk production didn't begin until the late 1800s. It was then that shipping by rail made the large-scale marketing of perishable foods practicable.

That's when dairy production became disconnected from local demand, and farmers began selecting breeding stock exclusively for their ability to produce milk. For dairy purposes, a well-fleshed body was not only *not* on the list of desirable qualities, it was actually considered a fault. The commercial imperative was to turn feed into milk, not meat, so a dairy cow's every pound of surplus flesh represented two things: feed that had been converted into the wrong commodity, and

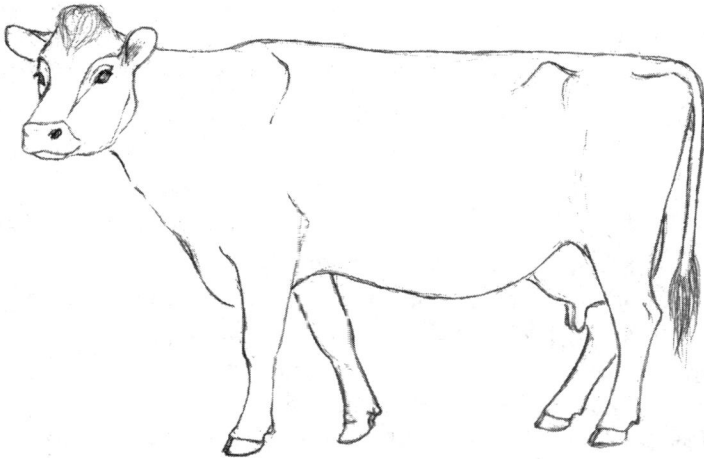

A homestead cow is designed for an active life on pasture, re-flected by a more muscular frame and smaller udder than those of her commercial dairy sisters.

flesh that had to be maintained by an ongoing caloric price tag. Over-night, the most desirable cow, from the standpoint of the dairy industry, became a skeletal carcass supporting a huge udder—great, if you're out to engineer the most efficient milk machine; not so good if your goals are animal health, resilience, and longevity. Beef breeds, meanwhile, were selected for size, muscle, and marbling. Breeding regimens have continued to follow those lines until the prevailing belief is that beef cows can't be used for dairy, and dairy breeds are unsuitable for beef.

Actually, the truth is much less black-and-white. Beef cows as a class are predisposed to turn food calories into muscle mass, while dairy cows, when lactating, will turn a disproportionate amount of their intake into milk, even to the extent of impoverishing their own bodies. But beef cows do make milk—many even make a lot of milk—and dairy breeds make delicious beef, even if they pack less on a carcass than their beef sisters. There are still a great many breeds that would be considered "dual purpose" for beef and dairy; and should the demand return, just about any bovine could be trained to pull an implement. So being aware that the breed of a cow is only a clue to her possible usefulness, not a restriction on the uses to which you might put her, should be prerequisite to selecting a family cow.

Regardless of the color of her hide or the heading on her registration papers (if any), the best homestead dairy cow is the one you can get your hands on—the one that is there. That said, you want certain traits in any cow you're planning on milking. You're looking for a cow that will hold still when you squeeze her; one that will tolerate your close proximity and handling; one that will make more milk than her calf needs. Without a doubt, in your search for these qualities, you'll be looking at some dairy breeds. But if you plan to raise up your cow's male offspring for beef, you may want a cow with a more muscular carcass, too, and this will incline you toward the all-purpose heritage breeds, or a dairy–beef cross. So breed is, at best, of secondary importance; individual qualities are going to count for more.

That said, breed can matter. Scottish Highland cows are cute, peering out from behind their bangs like Disney princesses, but all that hair is a liability in hot Florida. Zebus are said to be homozygous for A2 beta-casein (a milk protein some folks consider desirable), but if your winters are extremely cold, a breed from India might not be your best bet. If you want to raise miniature anything, more power to you, but remember that if your neighbor's standard-size bull gets over the fence and breeds your tiny cow, there may be problems come calving time.

Rather than attempting to select a breed like an online shopper looking for the best brand of running shoes, list the traits you consider desirable in a dairy cow, and then look around to see if you can find a good fit. Chances are, when you've found the cow that best satisfies your requirements, breed is going to look like a side issue.

Shopping

The first step is to see what's out there.

Shopping for a dairy cow starts with knowing where to look. Neither Walmart nor Amazon has yet figured out how to cut into this market; neither is there a used cow lot on the outskirts of town. Nevertheless, individual dairy cows are changing hands in greater numbers now than any time in the last sixty years or more. With the family cow making this stellar comeback, it's possible you may have a neighbor with a cow to sell.

And that's great, because our ideal place to find a cow would be in our own ecological setting. A cow that is used to the regional climate and plant species is going to have an advantage over an import. Local sales are easier—and less expensive—to get home, and if you can buy local, it will mean you'll know at least one neighbor who is also crazy enough to keep a homestead dairy cow. We'd like to find her on a farm that manages cows the way we intend to: all-grass-fed, holistically grazed, no meds or chemicals, hand-milked. If such a neighbor had a spare cow to sell, we'd be interested.

If we didn't already know where to find such folks, the local vet, the guy at the feed mill or grain elevator, or the rural postal delivery person might be able to give us a lead. Many small-town businesses keep a bulletin board for local transactions—ours do—where you can find firewood, ponies, puppies, and, sometimes, a spare dairy cow. The regional farm and dairy newspaper, if there is one, could be helpful, too, although ours mostly offers dairy cows by the herd, not the head.

If we couldn't find a cow by asking around, our next step would be to look online. Believe it or not, Craigslist often has individual dairy cows offered, and some of these come from a homestead setting. You want to know as much as possible about the animal's history, and the more information the seller offers, the more likely it is she has a cow worth looking at. Very brief descriptions often indicate a dealer who picks up cheap cows wherever and immediately resells; you could get a good cow this way, but it will be the luck of the draw. Keeping your search region tight will help you find a cow already at home in your ecological setting. Online chat forums may be a good place for inquiry, especially if they are regional; but many have national or international reach, and with these you'll have more sorting to do to keep your search local.

You may be tempted to try your local livestock auction, where it is not uncommon to see individual dairy cows or, more often, whole herds offered for sale, but we don't recommend taking this route. While you might find a good cow this way, there's no guarantee; you're looking for too many qualities that would be difficult or impossible to assess with a cow in the sale ring. In any case, the auction barn is for quick liquidation of generic or lower-quality stock; the most desirable animals are marketed individually, so their merits can bring the higher price they deserve.

Cow Qualities

How do you judge a cow's fitness when you have no experience? Here are sixteen things we take into consideration when shopping for a dairy cow.

Teat Length and Udder Conformation

There are a lot of cow qualities that are important to us, but this one comes first; if the working parts aren't right, there's no point in looking further.

Teats need to be long enough to be grasped with thumb and at least three fingers; the tiny, short teats the commercial dairy world prefers are a disaster for hand milking, and they don't serve for calves, either. You don't want them to be too big, though; in an old cow, teats can sometimes be so large it's hard to close your grip around the top. The size of your thumb is about right.

An udder has four quarters, each of which makes milk independently of the others and is drained by its associated teat. Ideally, all four of these are in working order, but sometimes a cow loses function

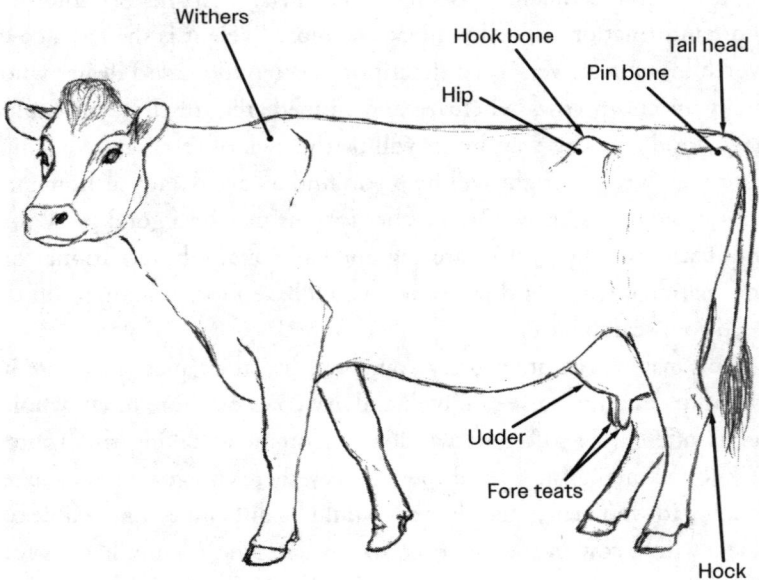

Some basic cow anatomy terms.

in a quarter; we call her a "three-titted" cow. By itself a nonfunctional quarter is not a problem for the grass-fed cow—her milk production should not be significantly reduced, because under natural levels of production the remaining three quarters can take up the slack—but you'll want to get some information as to how she lost function in that teat. Trauma, such as having been suckled by other calves when she was young, or being stepped on by another cow, doesn't point to an endemic issue, but if she lost function due to mastitis (an udder infection), you'll want to know more. If she's prone to mastitis, or has mastitis now, this will become your problem, and maybe you'd rather not start with the field slanted against you.

You'll sometimes come across a cow with an extra teat or teats, usu-ally on a back quarter. In a commercial dairy, these are clipped off at birth, but homesteaders may leave them (we do). They're often blind (nonfunctional). But sometimes you'll get a double teat—a single teat with two openings, one in the side. A nonfunctional extra teat, or auxiliary teat, is no problem, but the double teat is messy and inconve-nient, and would be reason for us to reject the cow.

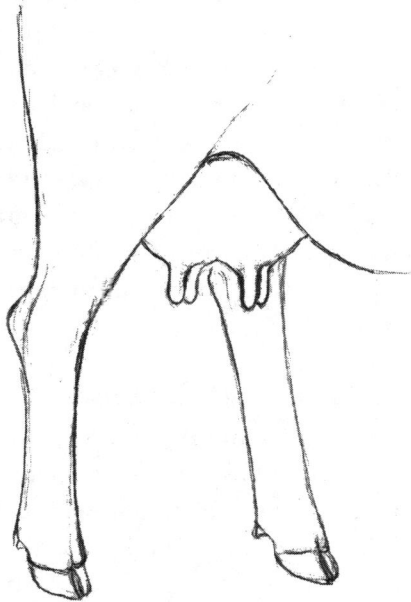

Our ideal teat is about the size of our thumb.

A good udder is large enough to look roomy, but not enormous. Commercial dairy cows have udders like sofa cushions, so large they have trouble walking or lying down. These cows are intended to produce 8 or more gallons (30 L) a day, not the 2–5 gallons (7.5–19 L) that are more common for a cow on grass. A grass cow needs an udder that will hold metabolically appropriate amounts of milk, and she needs to be able to walk without impediment. Her udder should be strongly supported, with plenty of room to get a bucket underneath. A well-hung udder looks graceful, not ungainly.

Diet

What has she been eating? In our perfect world, we'll buy a cow that has never had anything but grass or dry, cured grass (hay)—no grain, no silage (fermented forage), no alfalfa pellets, beet pulp, or fancy bagged probiotic fermented forage booster of any kind. Why? Because grass is what cows are designed to eat; furthermore, our land doesn't grow any of the other things, and we are wanting to fit this cow into our ecosystem, not fuel her with purchased, off-farm inputs. If this is our goal, we're better off starting with a cow that has received few or no commercial supplements.

That's *our* ideal. Many folks don't have any problem with feeding supplements—mostly with the idea of increasing or maintaining milk supply—and for them a cow with a previous diet of concentrates doesn't seem problematic. Whatever your preference, just know that you'll stack the deck in your favor if your cow already has the dietary habits you intend her to have. While we have transitioned a good many cows from a partial dependence on grain to an all-grass diet, switching a higher-production animal used to a grain-heavy diet to an all-pasture regime is probably best not undertaken by the wholly inexperienced; too much risk for her and for you.

Medication and Chemical Exposure

When we're shopping for an animal that will be intimately involved in our own diets, naturally we'd like to find one carrying the fewest possible chemical and pharmaceutical residues. Not only that, an animal that has received a minimum of pharma interventions is more likely to have the kind of natural, robust health we look for in livestock. That said, if a cow had most points in her favor, we probably wouldn't turn her away because she'd been vaccinated for blackleg as a calf, or been wormed at some time in her past; first, because there aren't so many dairy cows offered for sale that we could confidently hold out for one that fit our med-free model perfectly; second, because our environment—everyone's environment—is already saturated with chemicals, including runaway concentrations of glyphosate, antibiotics, and hormones, and a grass-fed dairy cow that might have some meds lurking in her somewhere is still going to be better security for safe clean food than anything we could buy at the store. Statistically, she's just a better risk. Better some cow, than no cow.

Prior Management

In addition to what she has been eating, and what or whether drugs have been in her past, we'd like to find a cow that recognizes and respects temporary electric fencing. Ideally, she's been under intensive holistic grass management—"rotational grazing"—and is used to daily paddock moves and the proximity of the grazier. If she's not used to these things, we can certainly teach her, but in our perfect world, she's already fence-trained.

Attitude

We're proposing to snuggle up to this cow once or twice a day for a length of years. How is she to handle? Is she comfortable around human beings? Find out. Stand on her right side, next to her belly and facing her tail; that gets you out of the target zone in case she throws a foot. Now run your hand along her spine and down her back leg to her udder. Does she fire a hoof at you? We don't buy cows that kick; it's no fun working with them, and you can't always train it out of them. If she's not tied or confined by a head gate, or "stanchion," does she let you approach her? It's OK if she moves away when you touch her—after all, you've only just been introduced—but she shouldn't be jumpy. Dairy cows are comfortable with humans; they *like* to be milked. Anything else isn't a dairy cow, she's just a cow that happens to be lactating.

Milking History

We don't use milking machines, and we prefer to buy cows that have not been machine milked; vacuum pumps cause teat and udder damage. If the cow you are looking at has been machine milked, examine the ends of her teats—the *os*, or teat aperture, should be smooth and tidy-looking. If it protrudes beyond the tip of the teat, or has a little sunburst of scaly protrusions, the cow is exhibiting hyperkeratosis—damage from machine milking—which will increase her tendency to contract mastitis. We also feel that machine-milked cows tend to have less vigorous let-down reflexes. Again, while this is not necessarily a deal-breaker, knowing her condition will help you assess this cow's value to you.

A related question is whether she has a nursing calf that comes with her. For the beginner, the calf can be a pretty big bonus. It does represent further pressure on the purchaser's pasture, so if space is tight you may not want the calf; ditto if it is a bull (or steer) and you're a vegetarian. And the calf won't cost nothing. But a nursing calf can be insurance against lowered milk production and mastitis during a novice milker's learning period, so it's something to think about.

Breeding Status

Is she bred? Even if she's already in milk, this is a significant question. A dairy cow is only a dairy cow if she makes milk, and she'll only make milk if she calves. If you're starting out with no experience identifying when a cow is in heat and getting her bred, it might be nice if someone else has already done that job for you.

If the owners say she's been bred, ask if they have evidence, in the form of a veterinary or lab report, that she "settled" (conceived to the breeding). It is not enough for the owners to have seen the bull or tech do the job; neither bull nor artificial insemination scores every time. If she's not pregnant, ask why—one of the most common reasons to get rid of a dairy cow is that she failed to breed back. If getting her pregnant is a problem, don't make it yours. More on this in chapter 11, Breeding.

Condition

Condition has to do with her overall appearance. Her coat should be clean and glossy: fluffy and thick in winter, sleek in summer. Her tail should not be clumped with manure. Eyes bright, nose moist but not snotty, ears alert. While your first interview with her may keep her attention too occupied for rumination, when she's relaxed she should be chewing her cud. And although it is normal for dairy cows to show their bone structure—hips and short ribs especially—she should not be gaunt.

Body Points

Beyond condition, here are a few other points we take note of.

Horns. We prefer naturally hornless (genetically polled) cows. The
gentlest of bovines can put out your eye if she swings her head at
the wrong moment, and horns can be an issue for getting in and out
of the stanchion. Self-defense may be necessary for wildebeest and
springbok, but dairy cows don't usually roam the range with wolf
packs, and we've never met a dairy cow, with or without horns, that
wasn't the equal of a coyote—or a dog. We appreciate the viewpoint
that if Nature bestows horns on a cow, they must serve many good
functions, but for the safety of our farm, we prefer cows that Nature
has not so blessed.

   We don't insist on natural polling. Calves can be "disbudded"
(their horn buds removed) when they are small, and if the job is
done cleanly you can't tell they aren't naturally polled—at least,
we can't. We disbud calves ourselves, when necessary, and wouldn't
reject an otherwise appropriate cow for horned genetics. Dehorning
a grown cow is another story—it's a fairly traumatic surgery; we've
seen it ruin a good cow, and we wouldn't do it again.

Feet. A cow that harvests her calories out in the pasture needs good feet
for getting around. In a younger cow, that means smooth, straight
hoof walls that make a 45-degree angle with the ground. Older cows
may have settled a little (consequently, their hoof wall versus earth
angle would be flatter), but you don't want a long, turned up hoof
wall and a cow walking on her heels, a condition not uncommon
in cows coming from a commercial dairy setting. Legs should be
straight, not bowed or knock-kneed.

Size and conformation. The modern Holstein, the most common
dairy cow in the United States today, weighs 1,500 pounds (680 kg)
and is nearly 6 feet (1.8 m) at the withers. Despite her massive
weight she is a bony, skeletal animal—modern dairy breeding
equates "gaunt" with "efficient"—and she gives over 2,000 gallons
(7,570 L) of milk per year on average, which is almost 7 gallons
(26.5 L) per day for ten months. She is *not* a homestead dairy cow;
she is a commodity conversion unit appropriate to modern agribusiness goals, and is no more a homestead dairy cow than a semi is a
commuter car.

A good homestead dairy cow should be on the small side—smaller animals make more efficient use of food energy—and her frame should be light and neat, but well-fleshed, not bony. She should combine a long lactation with moderate production and good foraging abilities. She should be a hardy, healthy animal that can feed herself and her offspring and make some people food, all out of that abundant but hard-to-crack molecule, cellulose. Until a hundred years ago, most cattle fulfilled these requirements, and we can still find those older genetics, especially among the "heritage breeds," the ones that have remained un-"improved" by industrial ag and land-grant universities.

Breed

People often select a breed because it is fashionable, hence (they think) it will bring big money when there are offspring to sell. Scottish Highland cattle come to mind; also, mini-anything. Maybe sometimes this plan works out, but fashions come and go, while good, practical homestead cows come in many breeds. Searching high and low for a particular breed, and passing up good cows that might otherwise serve you just fine, is likely to mean a long search and a high price tag when you find your cow.

Other reasons for preferring one breed to another may be more significant. Our Jersey–Dexter cross cows give more butterfat per gallon than our neighbor's Jersey–Limousins, a significant advantage for a household dairy cow. Dexters, or any dual- or tri-purpose breed, carry more beef on a carcass than would a straight dairy breed. Certain breeds are more likely to produce A2 milk than others, and so on.

A2 Beta-Casein Genetics

Which brings up the issue of milk proteins, specifically beta-casein, and the much-discussed relative values of its A2 and A1 variants, considered by some folks to have significant health and nutritive properties. This topic is under hot debate, and if it interests you, there is plenty of journalism available on the subject. If you think you have a good reason for preferring a homozygous A2 cow (one that produces no A1 beta-casein)—a reason that is worth paying for—be sure to get documentation of the status of your cow. This can take the form of genetic test results, or, should she be a registered animal, it may be noted on her registration papers. If you buy her, these papers will be yours, so ask to see them. The seller's personal assurance, however earnest, is not enough to justify the boost in price tag a homozygous A2 status represents.

Age

There are a number of permutations: If you want an easier transition to your dairying life, a cow already trained to hand milking has advantages over a heifer you'll have to train yourself. She'll be comfortable in the stanchion, and accustomed to letting down her milk for a human being. The calving history of the older, tested animal will be a known factor, and her milk records, if there are any, will give you a good idea of her productive capacity.

Younger cows, obviously, have more time ahead of them. Although commercial dairy cows are sent to slaughter at four years of age on average, an all-grass homestead dairy cow might live ten or more years—perhaps many more—so even if you buy a cow six or seven years old, she is likely to have several good years left in her. We ourselves sold a nine-year-old cow to Wild Sycamore Farm that is still giving them calves and milk four years later.

Hidden Issues

Why is this cow for sale? Ask the owner, and use your head. The owner doesn't want this cow, and there is a reason. It may be as simple as herd reduction—after all, on a farm where cows are having babies every year, eventually there will be more cows than grass—but do inquire. A previous miscarriage, failure to settle to breeding, or a history of mastitis are all good reasons to cull a cow, and while the owners might feel no compulsion to volunteer information about such problems, they may still value their reputation too much to conceal a flaw when asked outright.

Location

Where is this cow right now? The nearer she is, the less difficulty and expense in shipping her. Furthermore, a local cow will, or at least should, be familiar with local forages. And if the cow you are looking at is close at hand, you'll feel less pressure to make up your mind about her in a brief visit.

Do you like the farm she's presently living on? Do you like what you can see of their practices? Abundant, grassy pastures indicate good management; seas of hock-deep mud should make you wonder whether their animal husbandry is any better than their land-husbandry. You're going to take home with you all the free germs that will come with the dirt on the cow's feet and with the grass and manure in her gut; do you want them? Clean, healthy farms have good biota; barnyards with piles of manure and puddles of urine breed pathogens. Wash your boots when you leave, regardless.

Personal Dynamics

This is a people question. How do you feel about the people selling the cow? Do they seem honest? Friendly? Is this a hard sell? Never do business with a bully; you won't enjoy it at the time, and you'll carry those negative associations forward with you if you buy the cow. And nice people are more likely to have nice animals. Fact.

Price

What should you pay for her? You can come at this in several ways.

The first is her comparative cost: What are dairy cows going for right now? If the range is anywhere between $800 and $5,000, how do you know where she belongs on the continuum? While you might be very lucky and find an outstanding cow priced at the extreme bottom of the range, it's not all that likely. Well-trained, healthy dairy cows are usually owned by the people responsible for that good training and health, and those folks generally know the value of the animal and the work they've put into her. They certainly deserve to be paid for both, and, as the demand for family cows continues to go up, they're in a position to hold out for a good price.

Cows at the extreme apex of the price range, on the other hand, probably owe a good deal of the sticker price to something fashionable but nonessential, like the name of the farm she comes from, the skill of the marketer or web page designer, a momentarily popular breed or trait, or something striking in her appearance, like her unusual size, color, or coat length. Are these so attractive you're willing to pay extra for them?

Then there's her value to *you*, personally. If a cow can be had for $800, why would you pay more? Here are some reasons that occur to us. This is a purchase you're going to know personally and interact with intimately for years. In buying her, you will be launching into unknown waters, with plenty of obstacles that may be easier—or more difficult— to navigate depending on the individual qualities not only of the cow, but possibly of the seller as well. What dollar value might you place on a first year of dairying that is slanted in your favor by good training for the cow, or her outstanding individual excellence, or good mentorship for you? Would it be worth an extra dollar a day for the first year if your new cow was already bred? Would you pay a few hundred dollars more (the cost of a weekend seminar) if the seller offered to teach you to milk? Will the peace of mind of buying from a reputable, respected farmer instead of a dark horse, be worth paying for? Would it ease your anxieties in your first weeks of dairying? Our own opinion is that money paid to a good farmer for a good animal is money well spent.

Additionally, there's that lowest common denominator of bovines: what would she be worth as beef? Our son Thomas says that the low-end price for hamburger in his Philadelphia neighborhood at the moment is $13 per pound. Gasp. Use the local price for beef to help you make a rough calculation. If you bought this cow and changed your mind about her the next day, called in your buddies, and did a rough-and-ready slaughter, what value could you expect to get out of her?

At the bare minimum, if you slaughtered a lactating dairy-breed cow you might expect to get 25 percent of her weight in good beef, as well as a whole lot of excellent bits and pieces, soup bones, and so forth. So, to get a rough minimum for her value on a plate, divide this cow's weight by four, and multiply by the average cost of the beef you buy for your household. By this calculation, an 800-pound (365 kg) cow dressing out to (a bare minimum of) 200 pounds (90 kg) of high-quality protein, is worth, if hamburger is going for $10 per pound in your neighborhood, about $2,000. Two thousand dollars is middling for a family cow, and the figures we have given for meat harvest are at the extreme low end; a cow in good shape would be worth more, whether for dairy or for meat in the freezer.

Your good homestead dairy cow is probably going to be priced some-where in the middle—not quite a steal of a deal, but—oh, please—not boutique-priced, either. Ultimately, the best cow for your homestead is the one you can get your hands on, the one that is available. No mat-ter how perfect a prospective cow may appear, she'll have her ups and downs when you get her home; and if you buy a cow that turns out to have a lot of hidden flaws, there's at least one significant value to that: you're going to learn a whole lot, fast. We've done it both ways.

One thing we have *not* done, and would advise others not to do, is buy a cow sight unseen. It's easy to idealize in a description—every cow sounds pretty wonderful when described by a loving owner or a determined seller. But we've seen a good many folks who bought a pig in a poke and, come delivery day, got some unpleasant surprises. Mail-order cows mean you take an awful lot on faith. Every animal will come with some peculiarities; you're going to feel better about this cow's if you've made an informed decision about them. If she's worth the selling price, she's almost certainly worth a trip to see her first.

In general, leave your cash and your trailer at home when you go to inspect a cow. Don't buy on impulse; don't make rescue purchases. *Don't* fall in love. Keep the end goal in mind: a grazing, lactating bovine that will turn your volunteer crop of cellulose into high-quality nutrition. And then remember: the best cow is the one that is *there*.

Moving Your Cow

Before you can welcome her home, though, you have to get her moved. If you didn't buy her from the lady next door, and you don't own, or can't borrow, a stock trailer, you'll need help. Find out if the sellers will provide this service—they have a vested interest in your purchase—and be prepared to pay for it. If not, any rural area will have folks who will haul your animal for a fee. Hauling fees are usually calculated from the number of loaded miles: around here, right now, about a buck and a half a mile.

If you don't know someone who hauls livestock, check with the local slaughterhouse, if you have one, or your neighborhood large-animal veterinarian; for obvious reasons, both of these will almost certainly know a hauler. And, of course, if your new animal is small enough you may be able to move her yourself—we're not the only folks we know who have hauled calves in a van or the back of a pickup.

Try to keep her trip short. Before you decide it's OK to trailer your new cow halfway across the country, you might want to ride in one yourself. Luxury horse trailers do exist, but most livestock trailers are bumpy, windy, and loud. In cold weather, a stock trailer is a mobile freezer; in hot weather, it's a solar oven. In a crowded trailer, or one that is too small, animals may be unable to avoid bumping or rubbing against the side or door, resulting in bruises or open lesions. "Shipping sores" can be minor or more serious, but, regardless, imagine how frightened and disoriented your cow will be after a long trip in a tight box where each bump in the road is painful.

If you're hauling a cow for the first time, you may have some questions. We generally put a halter on a cow we are shipping and tie her head, making sure to give her adequate slack so that if she slips and falls her head isn't pulled up. We don't usually tie calves. Check every gate and door latch to make sure they are securely fastened, every time. If you can go through your life without the experience of seeing in your wing mirror the head of a steer attempting to liberate himself through the side door of your stock trailer as you barrel down the highway at seventy miles an hour, do it. Likewise, make sure the floor of your trailer is stout and has no spongy places; there are safer ways to trim a cow's hooves than grinding them against fast-moving pavement.

Before you can haul her, of course, you have to load her. Even with a loading chute this can be a job; in the absence of a chute or ramp that allows the cow to enter at trailer height, try backing the trailer up to a hillock or berm so she can load without a jump. Some cows will follow a lead rope into the trailer; some will load for hay or grain. When force is necessary, we generally use a long lead rope and pass it around something in the front of the stock trailer to give ourselves leverage. Then, with one or more person shoving at the back end, and another taking up the slack on the rope, we'll convince our reluctant cow that it is in her best interests to hop aboard. With yearlings as yet untrained, we find that if two of us use a bent stock panel to cramp them down into a small enough space, they'll get with the program and load themselves. With a bull, we use bait exclusively, never force; fortunately for us, all the bulls we have ever bought were trained to follow a bucket—the best use of grain feed we know of.

Single Cows and Herd-Mates

The present age is deeply concerned, and understandably so, with issues of justice, kindness, and nonexploitation. We applaud this. One idea that has gotten airtime lately is the notion that keeping just one cow—or pig or chicken—is a form of cruelty, founded on a failure to consider and respect the natural communal propensities of the species. Simply put, because cows (pigs, chickens) are herd (flock) animals whose wild relatives, in a natural setting, live in groups, it is suggested that if you keep a single cow (pig, chicken) on your small homestead, the animal will pine for lack of same-species company.

This is not a frivolous concern. Ignoring the nature of farm animals has been in large part how modern husbandry has gotten into such a mess in the first place. If our homesteading goal is to build healthy ecosystems in which human beings play a positive role, our success, if any, is going to be a direct result of the respect and reverence we show for the nature of Nature.

So, what *about* herd animals? Ought they only to be kept in groups? Certainly, wild ruminants, pigs, and chickens show strong grouping behaviors. But these populations are just that—wild—while their relatives with whom we are collaborating are *domesticated*, that is, they have been adapted over millennia of selective breeding to live in symbiosis with human beings. Over this long period our species have become mutually dependent. For most of this period—not excluding the present—this has included living in close association with, and even, in many cases, occupying the same dwellings as, our livestock. Humans and their animals have, in fact, become one herd.

And that's the answer to the one-cow dilemma. Your homestead dairy cow is *not*, in fact, deprived of her herd—*you and your family* are her herd-mates and primary associates. You visit her every day for an extended period of physical contact, grooming, and shared need. You participate in her migration over the pasture; you establish her daily territory; and, not you only, but all the animals on the farm, directly or indirectly part of your managed ecosystem, are her associates. Farm communities are not species-ist. As a matter of fact, most pet owners understand this principle implicitly, which is why we don't fret over dog owners who keep a single dog instead of a pack.

Nevertheless, expect that a cow coming from a herd situation to a single-cow farm will have to adjust to her new situation, and it's in both your interests to help her make a smooth transition.

At the outset, you'll want to make sure the first fences she encounters on your farm are convincing ones. We've seen folks move a new cow in

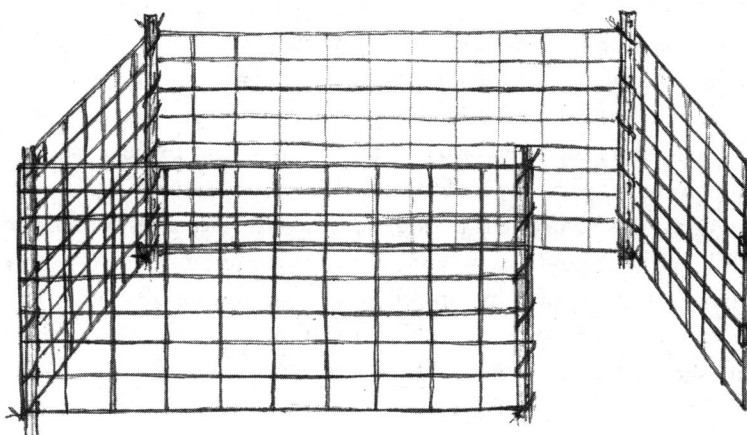

Stock panel corral.

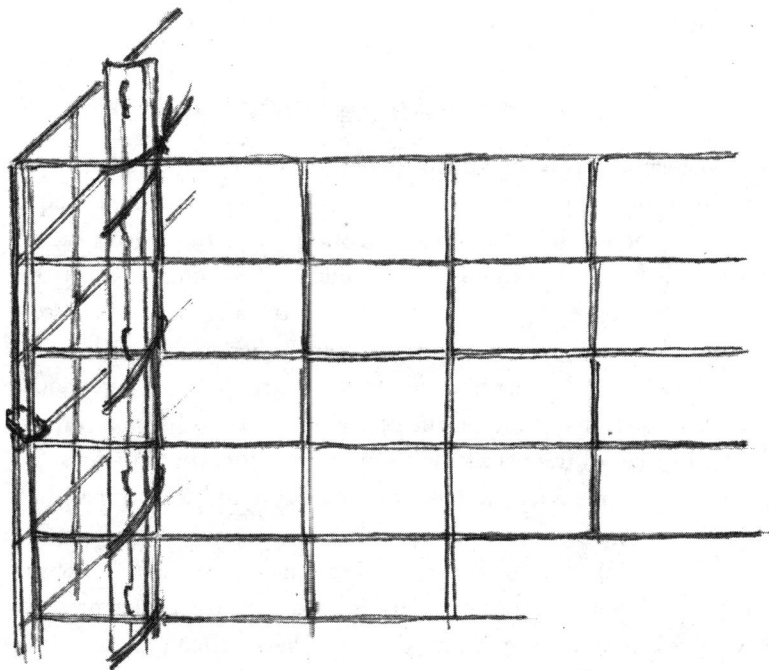

Close-up on the corner of a stock panel corral: t-post, stock panel, baling twine.

behind nothing but a single strand of polytwine—in fact, we've done it ourselves—but in general our advice is to put her behind a fence that isn't just a suggestion. If you don't have durable pasture fences or a stout corral, no problem: four welded stock panels and eight t-posts make a great temporary corral to contain your new animal for the first few days and prevent her wandering off the farm in search of something familiar.

During this adjustment period, you can help her gain confidence in her new surroundings by visiting her often, bringing treats of hay or fresh grass, salt, or kelp. Groom her, if she's comfortable with that. If you have purchased a cow with her calf, she's already got the company she most values, but she'll appreciate the near presence of your other farm animals. Watching a cow interact with the barn cat, or a stray chicken, will show you how alert, curious, and social a cow can be. The farm dog may not be the best welcoming committee; a cow that is familiar with hunting canines of any kind—dog, coyote, wolf—usually goes into defensive mode around a dog. Given time, she will teach even a nippy Border collie some respect, but keep an eye on him meanwhile and don't let him stir the pot.

Bovine Pasture Mates

If your new cow is being added to a herd—even a herd of one—you need to expect some hazing behavior. In socially more-or-less equal situations there will be shoving, mounting, even prolonged head-to-head trials of strength; and where animals of significantly different size, strength, or dominance are meeting for the first time, expect some flight and pursuit in the first encounter. Either way, if your fences are polytwine, make sure you allow some extra space for everyone to express herself. If you can let the early introductions happen through a corral fence, all the better; that way, even before the shoving starts, there have been some pheromones exchanged and the animals have some information about one another.

On the occasion of first turning the animals in together, make sure you've sized their paddock—the grazing space to which they presently have access—to include the needs of the new animal; an additional benefit to the added area is that they will have more room for boisterous behavior, as well as the welcome distraction of new forage. They're already feeling competitive; their competition shouldn't include food.

Beware of groupings in which some animals have horns while others have not. The only purported instance of death by toxic forage we've ever found convincing was in the case of a polled animal that was introduced to a herd of about a dozen horned cows. It was midwinter and there was only one round bale at which all the animals were supposed to feed; the only things growing in the pasture were nonforage plant species, some of which were high in silicates. The new cow was ostracized, as one would expect, but with the cows' round bale feeder located out of sight of the house and yard, no one really noticed whether she was allowed to eat. As the new girl on the farm and the only hornless animal in the pasture, she was at a double disadvantage for establishing herself in the pecking order. Even a smart cow, who would not under normal circumstances have the least inclination to take on a toxic load of an unfamiliar and probably unpalatable plant, might do so under the dual pressure of hunger and inhospitable pasture mates. This one did. To avoid unfortunate accidents, keep an eye on the situation; check in often; make sure all your animals are getting their share of good things.

And if your new cow is lactating, don't be surprised if her production is depressed for the first few days after the move. She's got a lot to get used to—different food, a different water source, different surroundings. Her let-down reflex, by which she ejects milk, is connected to her sense of safety and comfort, both of which will have been temporarily compromised. Take things slowly for the first few days; make sure milking time is relaxed and your presence is soothing. She's been through a lot; now is your chance to build her trust in this new relationship.

From Zero to Sixty

Mike and Heather Roberts,
Dove Mountain Cross Farm, Canton, Ohio

Already alert to the adulterated, compromised state of processed food, Mike and Heather Roberts were no strangers to food skepticism; with the 2020 coronavirus outbreak, they began trying to source their diets locally. "The food supply chain is fragile. We needed to put our family on a more secure, sustainable footing," says Mike. But when it came to producing food themselves, the family started out with no experience, and two major obstacles to their homesteading dreams: "money and fear."

Nevertheless, the family began researching food and food animal production on the homestead scale. Soon they bought thirteen acres of strip-mined land and dug in. "We went directly to the dairy cow," Mike says. "The dairy cow is the foundation of the homestead."

But according to their initial research, it looked as though the least expensive thing about getting a cow was going to be the animal itself. "Twenty thousand dollars to fence our seven acres of pasture; a $30,000 barn; and a tractor is a minimum of $15,000. How could we start out $65,000 in the hole?" Then, thanks to a couple of homestead workshops the couple learned that cows seldom require more shelter than shade, or the lee side of a barn, while small-scale portable electric fence can be assembled for just a few hundred dollars.

Searching locally, they found a cow at an Amish dairy. Low-production for commercial purposes, she was more than adequate for the Roberts' needs, and she had the two traits they most desired: gentleness and A2 milk. Furthermore, the price was right.

Fences were next. Instead of $20,000 for professionally installed high-tensile fence, the family built their own perimeter; and, for more durable containment, a corral of stock panels and wooden posts. Reels, step-in posts, and a charger for rotational grazing put the whole bill up near $5,000. "Our new cow was trained to respect temporary fence after only a week," the Roberts remember. For water, they used 500 feet (152 m) of garden hose; after a year, they knew where they wanted permanent waterlines, and installed a frost-free spigot. Shelter was even less of a problem. "We found that the cold wasn't really an issue. In our first winter, temperatures got down around 0°F (−18°C), but with a windbreak of stock panels and tarpaulin, the cows were perfectly fine."

Milking was similarly achievable. "All in all the first milking session took about an hour. Within a couple of weeks we were down to fifteen to twenty minutes. Tandem milking [employing two milkers at once] is the most underrated family time," says Mike, whose nine- and ten-year-olds consider it a competitive sport.

Three years in, the Roberts are firm in their belief that "even if a person has only limited funds, and no experience, with the right mentorship, research, and desire, homesteading is possible."

CHAPTER 5

Settling Your Cow In

Sooner or later it's going to happen—the cow that gets away.

Take us. Committed dairy cow people though we be, there was a time when we were inclined to think dairy steers didn't carry enough meat on their carcasses, so we bought four Herefords—two steers and two heifers—to add to our herd of a dozen or so Jersey-cross animals. We'd exercised due diligence: we made sure the new bovines were trained to respect electric fence, and had even, after a fashion, been under rotation. When we brought the animals home, we confined them in the closest thing we have to conventional fence, a small pasture enclosed by two strands of galvanized electric fence wire. We gave them time to settle in—and settle down, too.

These beef animals reinforced for us the differences between our Jerseys and the beef herds our fathers had kept. Unlike the soft dairy breeds, beef cows are made of muscle and endurance. Where an excited Jersey runs out of steam after trotting a few dozen yards, a Hereford might not slow down for several *miles*. The first few days went smoothly, though; everything seemed hunky-dory—until we undertook to introduce the new animals to the established herd. In retrospect—as in, like, five minutes into the situation—we recognized our mistake, but by then it was too late.

The Jersey herd was on the east side of the hill, and the pasture where the new cows had been acclimating was on the southwest side, out of sight and scent of the big herd. To bring them together, someone had to move. We have no permanent interior fence worthy of the name; when we move a herd, we can either set up a temporary lane with two reels of polytwine and some step-in posts, or we can set up what we call "wing fences"—a kind of funnel of temporary fence—point the cow or cows in the right direction, and drive them across the open field, in most cases arriving at point B with all the cows we set out with. The cows are accustomed to being moved in this way, and their long-established experience has been that, wherever we take them, they'll find good things—so, by and large, they're pretty cooperative.

In this case, our mistake was not in how we chose to move them; our mistake was in moving the Herefords at all. What we should have done (and have since done in every similar instance) was to take the calm, trusting cows to the fractious ones, not the other way around. After all, if anyone was going to cut loose—and whenever you move cows you know someone *may* cut loose—far better it should be one of our phlegmatic, acclimated, *slow* Jerseys than one of those fire-breathing, guided-muscle Herefords.

The irony is that if the new cows had just cooperated for another twenty yards, they'd have been within sight of the bigger herd, and that would have been enough to ensure they didn't run off. Not only are cows insatiably curious, they experience safety in numbers. These four, had they seen other cows nearby, would never have run in the other direction; more likely, they'd have put themselves away. Instead of which, almost within sight of our goal, they spooked and ran. The galvanized hotwire that makes the suggestion of a barrier between our pasture and the wide, wild woods was absolutely no obstacle for these excited animals. We spent the next two hours chivvying them up and down our steep Appalachian hollers. One never made it home, disappearing completely between Spring Run and Bates Branch; whatever herd he attached himself to, the owner never got around to calling the sheriff to report a stray steer. We hope he tasted good.

Which all goes to say, there's nothing like being prepared.

———

Many of us are prone to jumping into new situations without adequate preparation, and homesteading seems to attract such folks as honey attracts bees. We all know people who purchased chickens before they had a chicken house, or bought an impulse livestock guardian dog puppy. And it's usually OK. With small animals, at least, there's some forgiveness; if we can get our hands on a large cardboard box, everything will be all right. Cows are a little different; it's good to have some infrastructure in place before we bring them home. Not too much, though; so let's go through the list of must-haves and nonnecessities.

Infrastructure

First, the nonnecessities: six-strand high-tensile pasture fence, three strands charged; luxury barn for the cow to hang out in by day and dream in by night; in-ground pressurized waterlines and frost-free stock waterers; sixteen-station free-choice mineral feeder—and that's just to begin with. It's amazing how many things we, or the people who offer us advice, imagine that animals need, when in reality they are pretty low-maintenance.

Then there are the must-have items; let's take them one by one.

Fences

Your grass-powered homestead dairy cow needs the nutritional advantages of intensive, holistic grazing—read, "daily rotations"—which means breaking up your grassy areas into small paddocks with electrified temporary fence. In which case, the less permanent interior fence you have, the better, since this will let you custom-fit each temporary paddock to your grazing needs. Put permanent interior fence on hold for now, and don't clutter up the farm until you know where it ought to be—or if you need it at all. We have a book on fence and grass management for the homestead in the works now, where we'll be dealing with this subject at length.

Perimeter fence, on the other hand—the fence around the edges of your property (or around the parts you'll be actively managing)—is another matter. Of course it's desirable to have some kind of barrier, physical or at least psychological, that indicates to your livestock where you want them to stop. Perimeter fences come in many different shapes, though, and how ponderous they need to be is a function of what kind of livestock you are keeping, who your neighbors are, and how well you sleep at night.

If you're raising elk, 12-foot (3.5 m) chain-link fence is about all that will keep them contained, and they'll still find a way out sometimes. Goats, as goat-keepers will assure you grimly, can be confined by any kind of fence, so long as it's one that will also hold water. Our dads kept their beef herds in with four strands of barbed wire—most of the time. Fences are permeable; cows will patrol fences looking for gaps, so if there's a soft spot, they'll find it sooner or later.

But for a holistically grazed family dairy cow—a smallish, tame, frequently handled bovine—the elk fence is probably overkill, and even the barbed wire may be unnecessary. That's because your cow isn't going to have access to the perimeter fence, at least not very often or for very long. The fence she'll primarily be in contact with is electrified polytwine paddock fence, moved every day. Her (very brief) daily fence patrol will tell her that she's completely encircled with electric string, and once she's established that there are no holes, she'll settle down to grazing. When you move her, in twelve to twenty-four hours, she'll patrol the new fence, and so on. She isn't patrolling your perimeter fence, because she only encounters it on those rare occasions when she escapes her paddock.

Even if you incorporate a section of perimeter fence whenever your daily paddock takes you up to your property line, it will be just a short section, easily examined—by you—for possible security leaks. Before your cow can find and take advantage of sagging posts, broken wires, or trees on the line, you'll have found the breach and fixed it.

There's another reason why perimeter fences can be less substantial when you're practicing holistic grazing. In conventional grazing situations, where animals have unlimited access to the entire pasture, the forage inside the fence is degrading, while ungrazed areas outside the fence may be fresh and abundant. In this case, the grass really *is* greener on the other side of the fence—greener, and always out of reach. Not so in holistic management situations, where livestock move forward daily onto those greener pastures. Livestock accustomed to daily moves come to associate their well-being with paddock fences. They know that good things happen inside their polytwine enclosures, and have the confidence born of experience that, sure as the sun will rise every day, you will move fence and give them a good, nourishing,

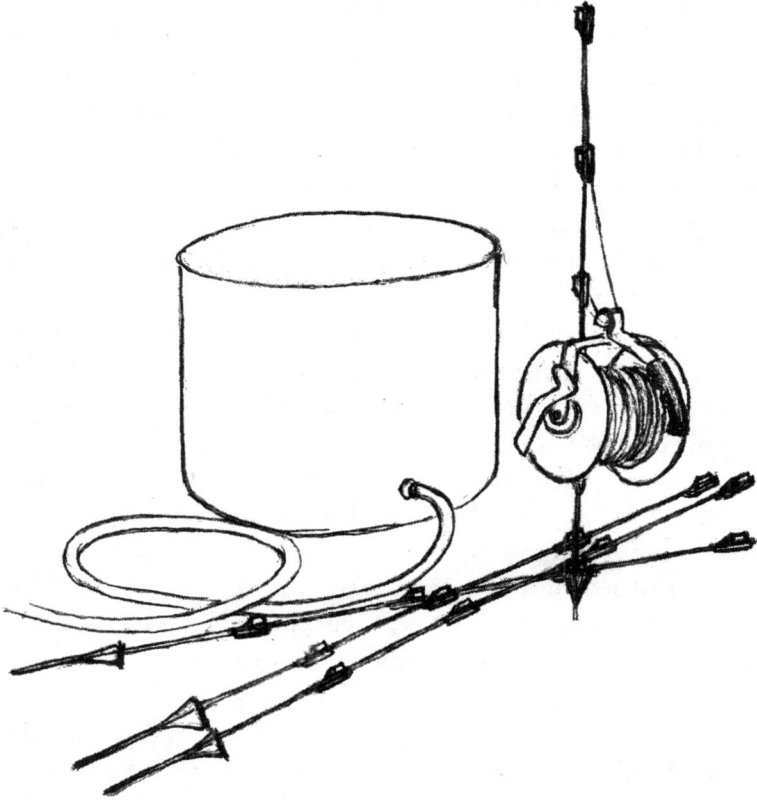

Simple rotational grazing equipment: reel and twine, posts, and a water tank.

belly-filling meal—so the pastures on the other side of the fence don't hold so much allure.

That's why your perimeter fence doesn't have to stand up to constant patrol by an animal looking for a way off the farm. Most of the time it won't be doing anything, because your cow will be inside her white string. On those occasions, however—and there will be those occasions—when deer have charged your temporary fence, or you neglect to turn on the charger for the twentieth time and your cow gets wise to it, or it's the first day of spring and she just gets bumptious—you do want some kind of perimeter fence, a visual indicator she'll understand that says, "I know you're on a spree, and you've got the whole farm to frolic in, but you don't want to go this way because there's a fence."

She'll get it. After all, once she's out, she's out, and she knows it. She's not looking to make a jailbreak, because she's already made one. After she kicks up her heels a bit she'll want to find some nice grass and settle down to a meal, and because you practice holistic grazing, the nicest grass in the neighborhood is right here, on *your* farm. After a while you'll look up from the kitchen sink or your laptop or the onions you're braiding and say, "My goodness, the cow is out. Darling," (this to your six-year-old) "will you go push her up toward her paddock while I figure out how she got loose?" And then you'll go straighten up a couple of fence posts, tighten up the lines, put the cow away, and turn the fence back on.

All other things being equal, anything that makes her hesitate at the border is all the perimeter fence your cow will need. All other things, however, are *not* always equal. There are two more considerations when you're designing perimeter fence: neighbors and sleep; or, in other words, how bad would it be if your cow did get loose, and how well-fortified against this eventuality do you need to be in order to sleep soundly at night?

If the other side of the fence is posted state forest land, we, personally, would be happy with a perimeter of a strand or two of galvanized hotwire. Our cow is not very likely to wander into the woods when the grazing (and footing) is better where she is, and even if she does, not much harm done.

On the other hand, if what's just over the wire is a six-lane interstate with unbroken lines of semis roaring by, it might be pretty disastrous if our black cows went for an illicit walk at midnight, so we'll probably sleep best with six strands of hot high-tensile—or maybe that 12-foot (3.5 m) elk fence—between our cows and the semis. Neighbors and sleep—definitely two of the considerations affecting what kind of perimeter fence you want to have. Just remember, permanent fence costs money, is labor-intensive to install and maintain, and you no doubt have many places to put your cash. Know why you are building a fence and what you need that fence to do, and be certain that you really want a permanent obstacle in that place, before you spend money putting a fence up.

A word about bells: they're cheap animal-escape insurance. If a cow gets out of the fence and out of sight, a bell will shorten your search-and-rescue operation by, on average, about a factor of six. Or, if you live in a forested area, or among crops or weed-choked fields, maybe sixty. It's a lot easier to find a cow you can hear, particularly after dark. Whenever we move cows or calves to a place they don't know, we put a collar and bell on at least one of them; it can make the difference between finding an animal in five minutes, or five weeks.

As for interior fences, you'll be able to take the best advantage of your forages if you keep permanent interior fence to a minimum. A good principle—and a traditional one—is to fence animals *out*, not in; that is, put permanent fences where they will protect sensitive areas like gardens, dooryards, springs, and so on, and leave the rest mostly open. That way, even when a paddock fence is breached, your tomatoes and prize dahlias are still safe. Also, this way you'll need a lot less fence, saving you time, labor, and materials.

Shelter

If there is one misconception about livestock that is more common than another, it is the idea that farm animals need cozy buildings to live in. They don't. They're animals, remember. None of their wild relatives ever had a barn, and they not only survived under these conditions, they throve. Yes, some animals, in some situations, will be better off with some shelter, some of the time, but you want to get it firmly in your mind that your new dairy cow is not a princess, and she doesn't need a palace. Keep your money in your pocket, and stop visiting that sales lot on the edge of town with all the cute prefab "Amish" outbuildings—you don't need them.

So, let's start with the assumption that your homestead cow is going to live outside, on pasture, most or all of the time. For what situations will you need some kind of shelter?

It's nice to milk under a roof when it's raining or snowing; it's nice to milk within four walls when there's a gale blowing. It's nice to let the cow carry the milk to the barn (milk weighs over 8 pounds to the gallon, or 4 kg to 4 L) rather than schlepping it up from the back forty yourself. It's good policy to keep some hay stored for—not a rainy day, but maybe a hurricane or blizzard. Rarely, there will be times when you'll want to give your cow more shelter than a tree line or the lee side of the garage. These are the circumstances for which we would design our homestead dairy barn, and it doesn't have to be fancy.

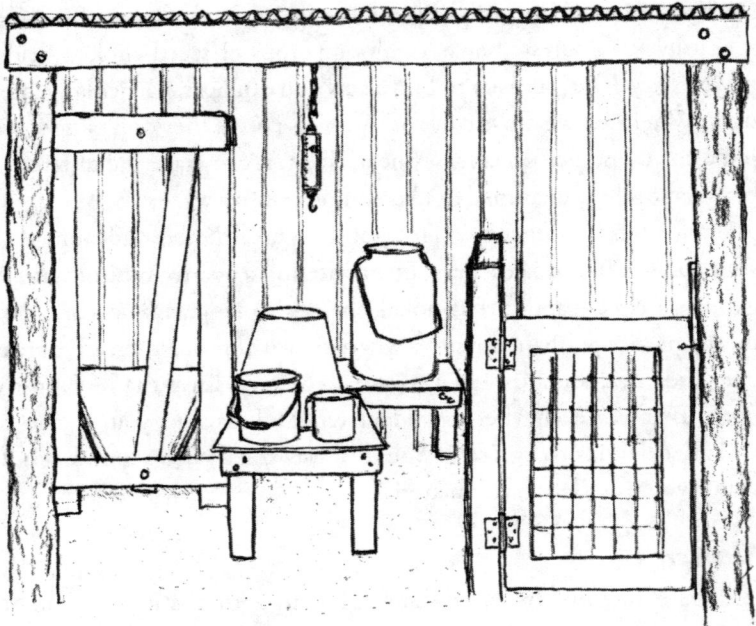

Everything you need—shelter from the elements, a milking stanchion, and a calf pen / hay storage area—in a small space.

Isabel, our introduction to dairy cows, was milked at various times in a chicken house, a three-sided shed, a portable canvas carport, a grape arbor (ineffective and buggy), and the field. Field milking is a perfectly reasonable proposition, in fact, for people who don't mind rain going down their collars (and dripping off the cow into the bucket) or snow settling on their heads while they milk. We found that after a while those things got old, however, so we mostly milk in a barn.

You can stack thirty square bales in about 5 × 10 feet of floor space, four bales high. A milking stanchion with plenty of room for cow and milker can fit in an even smaller area. Chances are you could adapt an existing structure for both purposes, or remodel 10 × 10 feet of your two-car garage. With a good barrier around the hay, you can use the same space to confine a calf, or for those rare occasions when your cow would actually be better off sleeping in the barn than in the field. We prefer an earthen floor in the dairy; packed hard and kept dry, dirt is more slip-proof than wood, and less likely to breed pathogens than concrete. Simple accommodations will get you started; later, if you decide to build something more opulent, you'll have a much better idea of what you want or need.

Water and Minerals

Water is heavy, and if you have to carry much of it, your daydreams will be about clever ways to move water that don't include you and a bucket. That's what hoses were invented for. The upshot is that when someone tells you the only sensible livestock watering system is buried, pressurized waterlines and permanent frost-free stock water stations, you want to believe them. Well, get over it. One or two cows don't drink that much, and on a small homestead you can follow paddocks with a 20-gallon (75 L) tank and a hose from the garden spigot. You can even carry water in a bucket without killing yourself. Remember, in-ground systems are more work to fix, and permanent water points, even if they are in the right place (they won't be) mud up and freeze— yes, even the frost-free variety.

Which is why we would advise you to skip all the expensive water systems. By all means, catch water from the house or barn roof and let gravity carry it to the pasture. Use a small watering tank you can dump without help—10–20 gallons (38–75 L) is adequate—and move it with the paddock. And while you're at it, skip the heavy multistation free-choice mineral feeders that are awkward to move and get left behind during paddock shifts and waste good money by getting rained on in the field. Find a cheap rubber pan that will hold a little loose mineral salt and kelp meal, and save your cash for more important things.

Tools—What You Really *Do* Need

If a dairy cow followed you home this afternoon and put herself away in the back pasture, you wouldn't have to rush out and buy a lot of equipment before the two of you could get down to business. We've tied a cow to a post, milked into a cooking pot, and strained through a clean jelly bag, and so have lots of other people, and it works just fine; and if that's your style, stick to it. But you're starting on a life journey of good dairying, so if you want a few dedicated tools for the purpose, here's what we would suggest.

A stanchion. This is a head gate for holding your cow in position while you milk, and you can make one with five salvaged two-by-fours and a few-odd nuts and bolts. A good stanchion is an important tool, and an inconvenient one is going to be inconvenient every day, so we go into what makes a good stanchion in chapter 6, Milking.

A milking machine. If your hands work, you probably don't need one. If you have osteoarthritis, though, get yourself the smallest, simplest unit you can find that has good reviews.

A small bench. Put it next to the milker, to hold things like the bucket.

A stainless steel bucket. You want the seamless kind with the rim turned out but not rolled. The bail handle should attach to lugs on the rim, not to those little welded jobs that come off. You want something durable, with nowhere for milk or grime to get lodged where your sponge can't reach it.

A stainless steel milk can or tote. Maybe; this is up to you. If you're milking more than one cow, you'll want a can to dump the first cow's milk into while you milk the second cow. The lid is nice for keeping out rain, flies, and anything the barn swallows drop.

A strip cup. This is a small cup with a screen top, and you shoot the first squirt from each teat through the screen, partly to get rid of any little thing on the teat aperture you may have failed to wash off, partly so if there is any flocculation (thickness) in the milk, you'll see it on the screen and investigate (more on this in chapter 12, Health, Unhealth, and End-of-Life Decisions). Actually, if you don't have a strip cup you can just shoot that first squirt against the side of your muck boot to accomplish the same goals. Likewise, you can use the screen without the cup.

A rag or brush. You want some way of cleaning your cow's udder before you begin. We carry a small bucket of water and a soft rag; some folks use a spray bottle instead of a bucket. Some folks use a soft brush, wet or dry. Dealer's choice. Most of the time, a grass-fed cow is going to have a clean udder, because with frequent pasture moves she'll seldom end up lying in her poos. Accidents do happen, though, so we wash, and in any case that preliminary lavage and massage become your cow's signal to let down her milk. It doesn't really matter what you do for udder cleaning, just be consistent; cows love routine.

Stainless steel bucket, milk can, strip cup, and wash bucket—
simple, durable, sanitary milking equipment.

Water. For washing the udder. Nothing else: no disinfectant, detergent, or such-like necessary; we're not trying to kill anything. More
in chapter 7, Milk Handling.

Something to sit on. This could be a three-legged stool, a 5-gallon
bucket, or anything else that will let you scoot in close under your
cow and hold a milk pail between your calves. We mostly use a
stool or a bucket, but when a cow with a really large udder freshens,
we've been known to milk for the first couple of weeks sitting on
the ground.

Kicking hobbles. Although you may never need them, it's good
insurance to have a pair of these restraints handy. A good dairy cow
doesn't want to kick you, but if something gets her started—the
barn cat spooks her, for instance—kickers will discourage her before
it becomes a hobby.

Making Everyone Comfortable

The first step to making your cow comfortable in her new setting is to relax and get comfortable yourself. Keep in mind that this is a dairy cow, which by definition means she likes being handled by humans, she's gentle, and she's cooperative. She's not looking for an opportunity to hurt you. Once you are comfortable with her, you can take steps to ensure she is comfortable, too.

Some things to consider: Water isn't just water, it's different everywhere, and she's not used to yours. It may take her a few days to trust your water source, and until then she probably won't drink much. Don't be surprised if she gives less milk during that time as well; her production should go back up when she decides it's safe to drink your water.

Food is a big factor in cow comfort, maybe the biggest. If she has plenty of good grass or hay, she'll spend more time eating and less time fretting about the change of setting. If she has new pasture mates, you want to make sure they, too, have food to distract them from the hazing they'll naturally subject her to. Hang around and watch; assure yourself that the established animals are letting the new cow eat and drink. A cow that is getting her belly full is in a good way to settling in.

If she is accustomed to a grain ration then, even if your goal is to eliminate grain from her diet, we suggest putting off her retraining for a while. If she was receiving some proprietary mix you can buy at the farm store, buy it; if you can't duplicate what she was eating, find the nearest equivalent and beg or buy a half-bag of whatever grain her last farm was giving her and mix the two to help her transition. Don't pile stress upon stress. Yes, grain is unnecessary for bovines, but she may not know that yet.

If she's in milk, make sure she gets milked the day you move her—don't overlook it in the bustle. Find out what her previous milking schedule was and how much she was giving. Ideally, have her milked before you load her, because after she's moved she may not let down readily. If you want to milk at a different time from what she's accustomed to, go ahead and start the new schedule right away, but don't leave a long gap; if she's used to a morning milking time and you want to switch to afternoon, do that first afternoon milking the same day as her last morning milking (twelve hours later), *not* the afternoon of the next day (thirty-six hours later). See chapter 9, Lactation, for more about designing a milking schedule. Once you've established the shift, be consistent; cows thrive and give more milk with a predictable routine.

It's a good idea to have a second person along at milking time for the first few days, in case your new cow doesn't like the looks of your stanchion. Milk her from the same side she's used to; if she's used to machine milking, you may want to have a machine handy in case she won't let down for your hands. One change at a time.

Security Matters

That first day on the farm you want your new cow behind some kind of hard fence, just while she's getting her bearings. Very soon she's going out on pasture behind, in all likelihood, a single strand of white string. Yes, it's electrified, and if she's trained to respect polytwine she should stay in all right, but there are a few things you can do to lower *your* tension factor in case she doesn't.

Bell her. A cow with a collar and bell is a whole lot easier to find on a dark night, or in the woods, than one without. Get one with a nice tone, not a clank; you're going to be hearing it a lot. We love our bells.

Use a halter and lead rope. Normally, safety is in favor of leaving a cow unhaltered when she's out on pasture—you wouldn't want her to snag on something and get hung up. But there are times when the safest thing isn't the best thing, and if you anticipate having trouble catching a cow or calf, a trailing lead rope certainly increases the odds you'll get her. If you find yourself chasing her cross-country, you're going to be really glad you left yourself a handle.

Tell your neighbors. This takes us to the other safety precaution for straying cows: neighbors. Why not let everyone in your immediate vicinity know the good news in advance and tell them you're bringing home a dairy cow? It's exciting, and those who aren't thrilled for you will enjoy the superiority of believing you're crazy. In any case, if you've given them due warning, they'll know who to call if they find a stray cow in their yard.

Bedtime is not the moment on which to dwell on straying cows, nor do you need to spend the first night in the corral with your new bovine partner to make sure she doesn't get the jitters. Yes, there may—will—be some unsettling moments ahead of you, but you've just taken a step that will change your life in a dozen ways that are profoundly good. So when night falls, check, if you must, to see that there's water in your cow's bucket and some hay within reach—and then go to bed, to dream of milk and cream and butter, and how lovely and green your pastures soon will be.

Working with What You've Got

Tim and Debbie Kasper,
Hiram College Farm, Hiram, Ohio

Granted permission to use ten acres of damaged, abandoned, degraded former farmland on the perimeter of a local college, Tim and Debbie Kasper, longtime residents of the area, didn't hesitate. Despite a heavy weed load and no outbuildings, restricted by a public footpath on one side and the college on the other, their success has still been notable.

"On a little piece of damaged land, one cow has provided hundreds of gallons of fresh milk, and a steady supply of yogurt, cream, butter, and cheeses, while our cow Noelle and our small flock of sheep have been turning land covered with poison ivy, multiflora rose, Canada goldenrod, and autumn olive, into pasture."

"Every week brings new lessons," says Debbie. "Once you've accidentally reeled a length of polytwine around the handle instead of the spool, you're highly motivated *never* to do it again. We're still hauling water in 5-gallon buckets—not a huge deal, but we can do better. And we've discovered we need a route on higher ground for when the cow has to get to the barn in wet weather."

Tim, remembering their first days with Noelle, is amused to realize how much comfort and delight he now takes in his daily chores. "If you want to make things happen, you have to take risks and not let fear hold you back; then things you were once afraid of can become things you love."

Family is at the center of the Kaspers' goals, successes, and challenges. "More than new skills or even physical injuries, the biggest hurdle has been differences in our modes of operation. This past year has brought them all into focus. Being more proactive in anticipating and working with those would have saved us some headaches—and heartaches."

Debbie concludes, "We have to stop comparing ourselves with others. Teddy Roosevelt was right: comparison really is the thief of joy. The best—and only—place to start is where you are, with what you've got."

CHAPTER 6

Milking

Everything has been building up to this moment: you're going to milk a cow, maybe for the very first time. Even if you've milked before and are good at it, there's a lot on the line with this life change. It's worth doing some planning to make your debut as successful as possible.

Where Will You Milk?

We've already talked a little about barn versus field milking. There are advantages to each. Field milking eliminates time spent driving or leading your cow to and from the barn; it eliminates the occasional barn cleaning; it may even eliminate the barn. On the other hand, if you milk in the field, it'll be you and not your cow carrying the milk wherever it needs to go; and weather, when it happens, will happen down the back of your collar. Barn milking means a stanchion, which means a cow that holds still while you are working; it happens in a place of your choosing; and it happens under a roof. Some folks want to combine the advantages of field milking with the advantages of stanchion milking by making a mobile stanchion that can be moved from paddock to paddock, an option which, we think, in addition to being cumbersome, eliminates most of the advantages of either system. Barn milking has worked well for us for over thirty years, so let's take a look at that.

Barn Milking: The Stanchion

In chapter 5, we pointed out that the barn in "barn milking" could be something as simple as a small shed or a corner of your carport. Banish from your mind all images of Dutch gambrel roofs and white ginger-bread cupolas; you could clear out the space under your deck and you'd be good to go.

The next thing you'll need is a stanchion, the most important part of which is a head gate designed to hold your cow more or less immobile while she is being milked. You'll sometimes see old metal stanchions at farm auctions, but they have no advantage over the simple ones you can make in thirty minutes with scrap lumber.

Building a Head Gate

Note the diagram for dimensions. You're making a 2- × 5-foot (0.6 × 1.5 m) rectangular frame out of two-by-somethings, with single upright side bars sandwiched between a double threshold and lintel at top and bottom, all screwed strongly together with at least two screws at each corner, so the frame won't wrack. That done, you'll want to measure 10 inches (25 cm) from the left-hand upright and drill straight through both the bottom bars with a bit of slightly larger diameter than the roughly ⅜- × 6-inch (0.95 × 15 cm) bolt you dug out of a can in the garage.

Next, back out the drill bit and slip a longish (around 6 foot, or 1.8 m) two-by-four (the "pivot") into the spaces between the top and bottom bars of the frame. Align it so one end is flush with the bottom edge of the bottom cross bars and the other end sticks out by several inches at the top. Drill again through the previous holes, passing through the sandwiched two-by-four; then insert your bolt through all three timbers, pinning them loosely together. Secure the bolt with an appropriately sized nut and a couple of washers, tightened only so much as will allow the pivot to swing freely.

Now, through the top bars, drill three holes—at 8, 10, and 12 inches (20, 25, and 30 cm) from the left-hand upright. Swing the pivot so that it passes between each set of holes, stopping to drill a hole in the pivot at each point. Find a bolt or whittle a pin that fits into the hole.

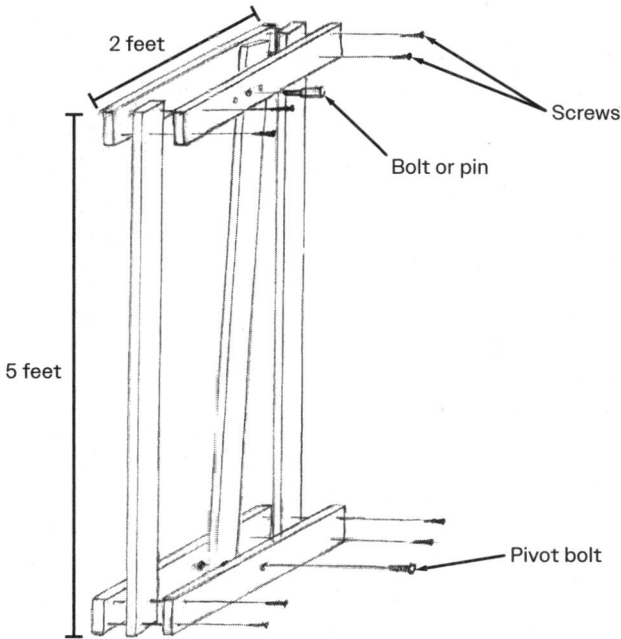

A basic head gate lets you put a stanchion anywhere.

See? Swung to its most open position, the pivot leaves a gap wide enough for a cow to stick her head into the stanchion between the pivot and the left-hand upright. To hold her there, you swing the pivot left to narrow the gap, and pin it in place while you milk. This arrangement gives the cow freedom to move her head up and down, but will keep her where you want her. The different holes let you accommodate cows with different size heads and necks.

Setting Up Your Stanchion

By itself, a head gate isn't adequate to immobilize your dairy cow. To make it a stanchion, you will need to attach it strongly to the barn wall, or a couple of posts, or whatever you choose that will keep it upright and leave room on the far side for your cow's head. It's customary to locate a manger there, which can hold hay or treats for the cow being milked, and it's most convenient if the manger has high sides so she can't push her food out of reach and then surge around straining after fallen whisps of hay until she puts her foot in the bucket. If you put the manger up off the floor by 8 inches (20 cm) or so, the barn cat can see that no rodent sets up housekeeping underneath.

The last requirement of a good stanchion is something—a wall, a horizontal bar, a well-placed post—that will prevent the cow swinging her hind quarters away from you while she is being milked. We milk from the cow's right, so our side bar is on the cow's left; if our stanchion was located in a corner of the barn, the barn wall would serve the same purpose just fine. Last winter, we set up a temporary stanchion in a neighbor's farm, just a head gate firmly tied to two t-posts with baling twine, with a third t-post to stop the cow pivoting; simple, easy to install, made of objects we had lying around, and taking a handy teenager about twenty minutes to construct. Voila.

The internet has somehow come to be dominated by images of a stanchion that is not only larger, more cumbersome, and more complicated than we've described, but has the even worse fault of being all wrong for hand-milked cows. If the computer images were labeled "stanchion for a cow milked by machine," fine and dandy, but they are not so labeled, they're ubiquitous, and homesteaders are building them by the train-car-load. You may have seen the one we mean: a big, framed box about 3 feet wide, 6 feet tall, and 8 feet long, on a heavily built platform several inches high, with a head gate and some kind of manger at one end and a horizontal board about 3 feet off the ground along either side to keep the cow caged in her box.

No, no, NO. Your cow's feet need to rest on the same surface as your stool—the ground—so you can scoot right up under her belly while you are milking. With her on a raised floor, your stool is held back by the floor of the stanchion, and you'll have to perch on the very edge in order to reach her teats. Further, the bar along her side is going to be right in your face, most inconvenient if you'd like to see what you're doing. We've seen people trying to hand milk cows in this kind of stanchion and it's an exercise in frustration as they try to hold bucket and teats at arm's length for the ten to twenty minutes minimum it's going to take them to milk in that position. Life's too short for such nonsense. Build the simplest stanchion you can, it'll last you the rest of your life.

Stanchion Training

Whatever kind of stanchion your cow was milked in on her old farm, it will take her only a few days to get used to yours. The issue will be

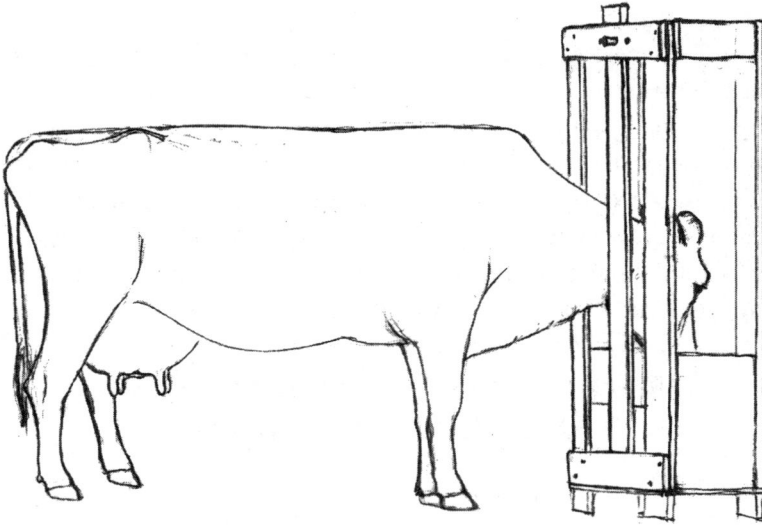

A cow secure and comfortable in a head gate.

getting her to put her head in until that happens. If you've bought a dry cow (she's not yet lactating) you have time to stanchion-train her before she calves, but if you bring her home Monday morning and you need to milk her Monday night, she'll need a crash course.

Pre-lactation stanchion training is easily done by offering treats in the manger, then, when your cow has put her head in and been trapped, brushing her, rubbing her down, handling her udder, and otherwise setting a precedent for future milking sessions. You don't have to put a lot of time into this. Give her access to the barn and leave treats in the manger so she gets used to putting her head in. After a few days, hang around until she's in and then close the head gate and let her struggle for a while. Offer more treats. Then repeat the process until she's used to being stanchioned and handled. Very convenient; we never do it.

If, instead, you, like us, have neglected to train your heifer, or the cow you just brought home doesn't like the looks of your stanchion and declines to put her head in under any persuasion, the situation is not dire. Get a halter on her, or a rope around her neck, and enlist the help of another intrepid homesteader, because you're going to use force.

This encounter, like so many that involve dairy cows, is just one more piece of evidence proving that dairy cows really like to be handled, because as big as they are, if they didn't like it, it wouldn't happen. Two people, no matter how buff and capable, are no match for a cow that really wants to be somewhere else. But watch this: you're going to slip a rope on your girl and, with some help from a stout pusher, get her head in the head gate. Clip your lead rope to the halter ring located under her chin, run it through the head gate, and then, pushing on the back end of the cow and taking up slack with the rope, leverage her into the stanchion. Don't give up; you're going to win this one, and once she's properly loaded, milking will be a cinch.

Milking Your Homestead Cow: Hands or Machine?

Squeezing cow teats is going to get your forearms really jacked and put gallons of dairy products on your table, and you'll be a practiced milker in just a couple of weeks. The first few days, though, will certainly have their challenges, and it won't be surprising if you begin thinking maybe the solution is to buy a milking machine. Resist the impulse. Milking machines are noisy, expensive, and subject to breakdown; furthermore, machine milking causes teat damage, reduces the quality of milk products, and inclines the cow to mastitis. If your hands are in working order, skip the milking machine.

The Downside of Mechanical Milking

Just in case you need it, here's a brief history of milking machines: no one needed them; farmers didn't ask for them. The handful of dairy cows belonging to the average farm could easily be hand-milked by the farm family. Then came railroads and refrigerator cars. Suddenly food could be shipped quickly and a long way without spoilage, and big cities got even bigger. With a demand for more food than the local environs could provide, farmers all over the country found themselves in competition with one another. On this national stage prices flattened out until low per-unit profits drove up production goals beyond what the land and farmer could accomplish without grain feeds, chemicals, and mechanical interventions—which includes things like milking machines.

So agribusiness was born, and milking machines made it onto the farm. Without financial pressure driving the shift, they would never have been adopted on a large scale, because they caused so many problems. That part hasn't changed; with today's enormous pipeline milkers, as with the early bucket milkers, the most noteworthy aspect of machine milking, after its admittedly fast service, is the damage it does to teats, udders, and the milk itself. Clinical mastitis, hyperkeratosis (damage to the teat orifice), and widespread milk contamination are all exclusively or overwhelmingly quality issues belonging to the machine-milking world. Significantly, it is this milk the USDA is testing when it condemns raw milk consumption wholesale: milk from bulk tanks, not milk from the hand-milker's bucket.

And, in case all this isn't enough to convince you to hand milk your new cow, there's the cost of buying a milking machine, the expense and trouble of upkeep, and the difficulty and time that go into keeping milking machines clean. Not that any of these aggravations is extreme (well, milking machines *are* expensive). It's just that we modern people tend to assume that if there's a machine to do a job, it ought not to be done by hand (or foot—think of people who mow postage-stamp lawns with riding mowers). But there are still lots of jobs, and milking is one of them, that may be done better, more conveniently, and less expensively by hand.

Note, however, that the value of home-produced, grass-produced milk is so great, we'd machine milk in a heartbeat if our hands stopped working. It's only *nonobligatory* machine milking we would encourage you to reconsider. All you folks who are using machines because for some reason you need them, hold your heads high; you're homestead heroes.

Back to bucket milking. If there's a downside to hand milking, it could be that it takes a few minutes longer to hand milk a cow (five or ten minutes, maybe)—only this is also observation-of-cow-well-being time, cow-massage time, and cow-friendship time, as well as say-a-prayer-of-thanksgiving-for-Nature's-bounty time—time well invested and possibly indispensable, in other words. Alternatively, someone might object that hand milking requires strength and endurance, and keeps you hard at work when you might instead be leaning on the compressor counting the minutes until the milking machine sucks dry; only, we don't mind having jacked forearms, and we do object to waiting on machines.

The Milking Stool

Milking stools come in all sorts of shapes and designs—three-legged, one-legged, wedge-shaped, strap-on—so it could be easy to get the impression that design is really significant. Well, so it is, as with any tool you are going to use over and over, every day, for years and years. Still, it's a need that can be filled very simply by just about anything that will support you securely at the appropriate height and let you scoot right up close to your cow while you are milking her. It needs to be of a height that lets you push your legs up under the cow; you're going to hold the bucket between your knees, or between the calves of your legs, *not* put it on the ground. You should be able to reach the cow's teats without hunching down, and to rest your forearms on your thighs while you are squeezing. Since humans and cows come in different sizes, it's not unlikely that you'll need different milking stools (or stool equivalents) for different people or different cows.

Still, lots of things can fulfill the necessary requirements, some of which you probably have lying around. A utility bucket is the most obvious and the most common on our farm; they come in several heights, and there is no shortage of them. We use them with the lid on, and right-way up, which we find more comfortable and easier to scoot around than upside down. We also use a three-legged stool that one of the boys made years ago—it's hickory, very sturdy, and the seat is roughly padded with an old cushion. If an older cow freshens very full and her udder hangs unusually low, we may use something much shorter for a few weeks, like a block of wood, and we've occasionally milked sitting on the floor.

The Milking Schedule

When is the right time to milk? And how often? The answers vary, but some points are consistent. Milking time, whenever you choose to set it, should be the same every day. Oh, sure, variations of a few minutes are fine, and the occasional hour-late milking isn't going to hurt your cow, either; but within reasonable limits, milking should be the anchor around which the rest of your day pivots.

If you're only milking once a day (that's the OAD you've seen on dairy cow forums), you can put that time wherever you'd like it. Mornings are nice, and get the day off to a good start; afternoons, last thing before dinner, let you call it a day when the dinner dishes are done. Lots of folks who milk OAD like a late morning milking time, and that works, too.

If you're milking twice a day (that's TAD, you got it) those milking times need to be evenly spaced, or as nearly as possible. For most of us that's going to mean a morning milking, not too late, so we can finish the day with an evening milking that's also not too late. In the beginning, we milked before breakfast and after dinner; when we wanted an earlier evening milking, we had to move the morning milking up correspondingly. Yes, that means getting up pretty early, but after twenty-five years we actually prefer it.

Pre-Milking Protocol

So, it's milking time, and your lovely new cow is in the stanchion. Your stool is handy, and your equipment—washing bucket, milk bucket, strip cup, towel—are on a conveniently located bench or shelf. Now's the time to develop good milking habits. Cows love routine, and a good dairy protocol will help you produce more milk, better milk, and a longer season of lactation.

Udder Washing

Udder washing has two purposes: to remove anything on the outside of the udder that might fall into your bucket of milk and cause inconvenience; and to stimulate the cow's let-down reflex so she'll give you the milk. We bring a small plastic bucket and a soft rag down to the dairy and use plain water—warm in winter, cool in summer—to give her a quick wipe-down. We keep a hand towel by the stanchion for drying. Often—in fact, usually, if you are moving your cow's paddock daily—she'll come in with a clean udder and there will be nothing to wash or wipe off. Wash her anyway; it will eliminate any stray hairs, flakes of hay, bits of dried manure, and it will make your cow happy.

Every once in a while your cow will lie down in a cow pie and you'll have to swill her udder to get it clean. No big deal—do your best, dry her off, and hope she doesn't make a habit of it, like our cow Poppy, who used to deliberately seek out warm, squishy cow pies to land her udder in. We don't recommend putting any kind of antiseptic or detergent in the wash water, since these can leave residues; your raw milk is naturally probiotic, and you don't want to kill any of those beneficial microbiota. The same goes for antimicrobial teat dips like betadine; protocols like that are for commercial dairies, where it's hard to keep cows, machines, and milk clean, *not* for the homestead dairy.

Once you're finished drying the udder, keep your towel handy—like over your knee or on your shoulder—for any extra drying or wiping up you may want to do. We find milking far more comfortable with dry hands and a dry udder.

Initial Stripping

With your cow in her stanchion, you on your stool or bucket, udder washing done, and a towel handy, you've still got one more thing to do before you reach for your milk bucket: you're going to *strip* your cow's teats. This means you're going to take the first couple of squirts from each teat, but you won't send them into the bucket; instead, they'll be directed into a strip cup. (A strip cup is just a cup with a bit of screen that fits over the top, like a grease can on the back of the stove.) This is by way of being our daily well-check or quality control for the cow and her milk. The screen is there so that if there is any thickening, or *flocculation*, of the milk, it'll stay on the screen and you'll see it and start investigating causes (see chapter 12, Health, Unhealth, and End-of-Life Decisions, for more on this topic).

Draw one or two squirts from each teat into the screen of the strip cup. *Now*, you are ready to milk.

The Milking Process

Scoot your stool right up to the cow. Stick your legs out straight under her belly, sitting as far toward her back end as you can go—it's OK to push her legs back a little with yours; in fact, you're entitled to if her feet are cramping your space. Try not to worry about getting kicked; first, she's *very* unlikely to want to express herself this way; second, if she decides to kick you, she'll probably do it too fast for you to see it coming, so what's the odds; third, getting kicked by a cow is more annoying than painful. It's the bucket you'll worry about, more than your body.

Position the bucket firmly between your knees or calves—you're *not* going to rest it on the ground, where it could easily be kicked over, stepped in, or spilled, and where it will pick up little presents off the barn floor that could drop into the milk later, when you're pouring

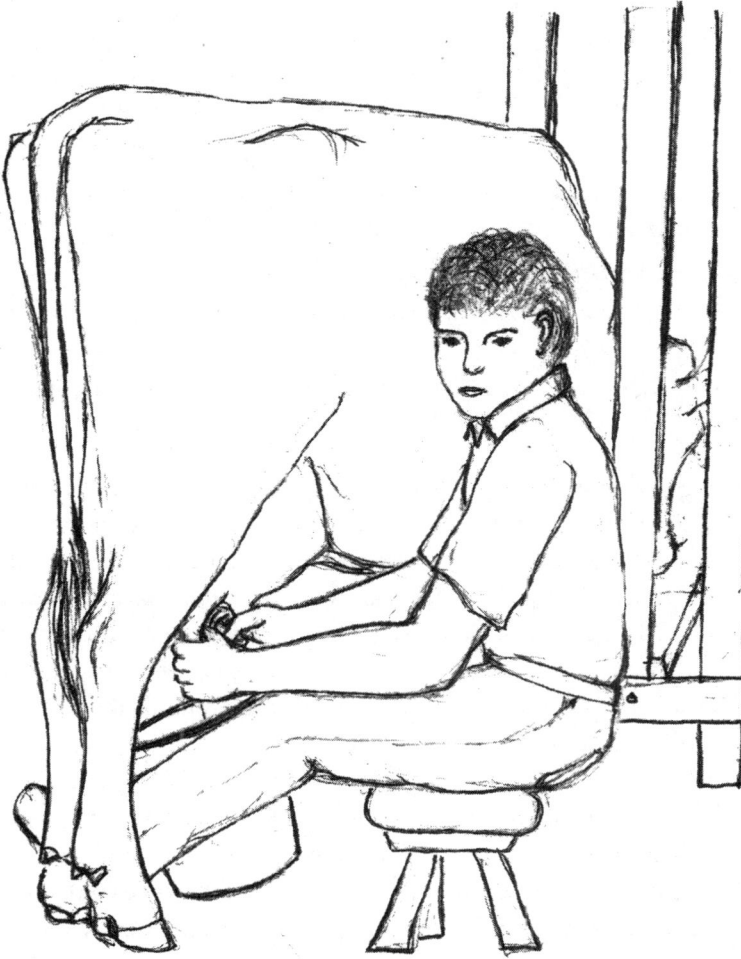

Comfortable, efficient milking requires snuggling up close to the cow.

from the bucket. Place the bucket up under the cow's udder, so that you can squirt straight down.

There's more than one way to skin a cat, as Shawn's father used to say, and there's more than one technique for taking milk out of a cow. In Europe, the norm seems to be to grasp the teat up high with two fingers and slide down, and they've been milking a long time. Presumably they know what they are about; still, we virtually never use this method,

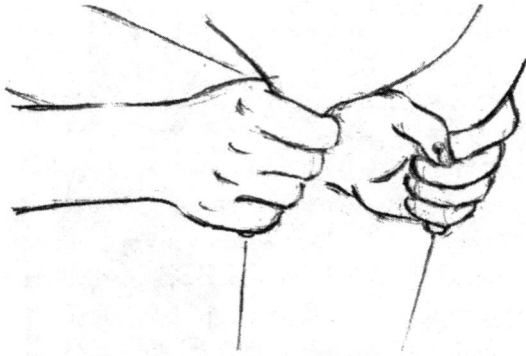

Teats should be long enough to allow you to get a good grip.

being of the school that thinks pulling teats is unnecessary, barbaric, and contributes to teat damage like hyperkeratosis. On our side is a lot of incidental literature that says pulling teats causes mastitis, as well as our own sense of touch, which tells us we're roughing up some delicate tissues. We are not teat-pullers. (Just to make sure you're confused, this way of drawing milk from a teat is called *stripping*, too, just like those first squeezes that clear each teat at the beginning of milking. One term; two meanings. Don't blame us, it wasn't our idea.)

When we milk, we use our fingers and palms to squeeze each squirt of milk out of the teat by closing off the top of the teat with (usually) the thumb and palm, and then closing the rest of our fingers, top to bottom, to force milk out of the teat. It's necessary to keep the top of the teat pinched closed, in order to prevent the milk simply squirting back up into the udder. When you do it right, you get a good, long squeeze of milk. That's the protocol—close off the top of the teat, then squeeze the rest, making sure the lowermost finger is the last to close (so you don't trap the milk in the teat).

Open your fingers again, making sure you release the uppermost digits so milk can flow again into the teat. Repeat the squeeze sequence: close the top of the teat and keep it closed while you squeeze the rest of the teat, bottom finger last. You're going to be squeezing hard, maybe much harder than you imagine is comfortable for the cow. No fear; her calf would squeeze a lot harder, and rough her up to boot. Eventually, you're going to do this two-handed, alternating hands, and pretty soon you'll have a good rhythm going: right, left, right, left, swish, swish,

swish, swish into the bucket. Your arms will get tired and your fore-arms will feel like sticks of wood and your enthusiasm will flag. Stay in there—milking will be a marathon for a week or so, and then, suddenly, it will be manageable.

Until she is trained to your routine, your cow may delay giving you a let-down; if that happens, your squeezing won't produce much right at first. You may have to massage her udder, try milking, massage again, try again, several times. If she has been milked before, she'll let down soon enough. If she's just dropped her first calf, give her time; she'll get the hang of it soon (for more on calving, early milk protocol, and calf sharing, see chapter 10, Calves and Calving). If she's late (six or seven months plus) into her lactation, she might feel empty initially; keep squeezing though, and her milk ejection reflex will kick in.

Keep this description in mind as you begin, but give yourself leave to experiment with variations. Cows differ; hands differ. Your cow may take a little while to let down. Learning to milk is like learning to ride a bicycle—explanations may be helpful, but only practice will get you home, so stay in there and see what works for you.

Final Stripping

Milking follows a general pattern. First there is the *let-down*, when you will see a good flow of milk. Actually, in a normal milking you'll get several let-downs, but if the milk is flowing well, you probably won't notice the subsequent ones. Then, as the udder is drained, squirts become smaller and the milk chamber begins to feel empty.

This is the time to switch to milking each quarter with two hands: one to squeeze the teat, one to manipulate the quarter, massaging to encourage complete milk ejection. Go round and milk each quarter successively until you get no more milk; then go around again. Keep repeating this, stopping periodically to press and rub the whole udder between your hands, then milking each teat, until you get almost noth-ing at each squeeze. You won't get nothing at all, because the udder is making milk all the time, but when you get only a tiny squirt, or just droplets, you're done.

You don't want to omit this last step, for several reasons. If you don't take all the milk your cow is making, she'll make less—the bovine version of the law of demand and supply. Further, incomplete milking can cause mastitis, and you don't want that. Finally, the last milk out of the udder contains the bulk of the butterfat, and you want every bit of beautiful cream for your tea and coffee, for butter making, and for sour cream, cream cheese, crème fraîche, and all the other lovely, high-fat, delicious, healthful dairy foods you're going to make. So, keep squeezing until the font runs dry.

Post-Milking Protocol

When you can't get anything more than a few drops at a squeeze, you're done milking. Our next step is to rub some bag balm (udder balm) into the cow's teats. Some folks skip this step; not us. Normal wear and tear on teats can leave them torn up from calf teeth, thorny pasture plants, and chapping. Not only is this uncomfortable for the cow, it can be a vector for infection, and any pain it causes the cow during milking may be an incitement to kick. Soft, pliable teats give a more satisfactory experience for the milker, too. Seems like all the virtue is on the side of using bag balm, so we do.

What kind of bag balm you use is, of course, up to you. You can almost certainly find some nice local person who makes a good salve or balm with fats you can feel happy about. Tallow balm, for obvious reasons, recommends itself as appropriate. We like a little oil of clove in our bag balm; likely you can think of some healing, soothing inclusions of your own.

Before you proceed, though, you want to set your precious milk bucket somewhere safe. In our dairy, each milker has a bench on his or her right; this gives us a place to put milk bucket, wash bucket, bag balm, towel, and so on. With the milk bucket securely on the bench, rub a bit of bag balm into each teat; you don't have to slather it, a little will do. Rub the rest into your hands.

If you are keeping production records on your cow, and there are good reasons to do so, this is the time to weigh and record your milk yield. We keep a binder in the dairy where we record each cow's name, production, the milker's name, weather conditions, and any random fact that we feel might be significant. This is where you want to write down things like "saw (cow's name) mounting the pasture pony" or "signs of mastitis, right front quarter." An ongoing record of such facts will be helpful for times like when you're trying to diagnose a trend

Day	Date	Temp	Milker	Cow	Yield	AM/PM	Notes
Sunday	11/24	41°	A	Pansy	5.5	A	Rain ½"
			M	Prim	7.5	A	
		52°	D	Pansy	6	P	Pansy in heat—sticker scratched
			M	Prim	8	P	
Monday	11/25	38°	M	Prim	6	A	Pansy A.I. COBA 'Bowtie'
			A	Pansy	7	A	
		52°	D	Pansy	7	P	

Recording milk production.

("When did Daisy's milk production peak?") or focus in on a calving date ("Note: bull riding Marigold June 14").

If the milk is going up to the house in the bucket, make sure you cover it with a lid or towel to prevent inclusions from the barn swallows. We carry milk up to the house in a can, mostly because it helps us avoid spills. If there is to be a delay before you get the milk to the house, you may elect to strain your milk now, in the barn; the less time that stray cow hair, drowned fly, hay flake, or manure dust (yes, really) spends in the milk, the sweeter the milk will be.

If the water we used for washing udders is still relatively clean, we use it to rinse the (now) empty milk bucket; removing residues right away means that, up at the house, it will be easier to get the equipment perfectly clean.

Teat Dips

In the commercial dairy world, it is standard practice to dip teats, before and after milking, in an antiseptic or disinfectant solution. The idea is to increase milk purity—prior to milking, by killing germs on the outside of the teat (which germs, as the teat will be bathed in milk during machine milking, would be included in the milk harvest); and post-milking, to decrease the rate of mastitis in the herd by killing germs that might enter the teat via damaged teat apertures. Given the conditions of most commercial dairies—large herds in confinement, concentrations of urine and feces on concrete, no purifying sunlight and wind—this is probably a reasonable protocol.

On the homestead, though, we consider teat dips unnecessary at best, and a disruption of the natural biological balance at worst. Unnecessary, because with hand milking, the milk never has contact with the outside of the teat; and even if the homesteader is using a milking machine, the likelihood of contamination while milking one grass-fed cow is far less than with one thousand cows on concrete, because the environment is cleaner to begin with.

A more significant negative to antimicrobials in this application is that they kill microbes indiscriminately. A cow's teats are home to millions of probiotic lactobacilli, the very microbes we want in our milk, in our dairy products, and in our guts. These bacteria colonize milk, and it is their presence that protects it from pathogens. In commercial dairies, where it must be assumed that every cow's teats may be carrying millions of germs that could spoil milk or cause disease in humans, the benefits of using antimicrobial teat dips to kill these pathogens probably outweigh the unfortunate side effect of also killing beneficial lactobacilli. In the home dairy, and in holistic grazing situations, where there is no reason to assume the presence of pathogenic bacteria, teat dips simply compromise the protective bacteria without providing any offsetting benefits. Hence, we don't use teat dips.

Treats or No Treats?

We don't typically offer treats at milking, and people frequently ask, "How do you get a cow to stand still without treats?" Contrary to expectation, cows don't require feed incentives to stand still for milking. They *like* to be milked. It's a pleasant massage for them, a personal-wellness moment, if you like. All you mamas who have nursed babies know what we are talking about. Also, dairy cows are in the habit of being milked, and they like habits.

Treats can be useful while you are training your cow, of course, but become a problem when they are withheld or not available. It's a question of expectations; if the cow is used to getting treats, she'll probably be naughty the first few times she goes without, just to let you know what she thinks; then she'll settle down into the new routine. When you are withdrawing treats, gradually reducing the amount offered might help to avoid tantrums, or it might just prolong the agony; cold-turkey abstinence establishes the new normal immediately, and your cow will soon get used to the change in routine. However you decide to manage, don't let a cow's misbehavior drive the choice; *you're* in charge; *you* decide how you want things to be.

When a Cow Kicks

We've said it many times, and we stand by it: dairy cows *like* being milked. Big as they are, if they didn't like it, it's hard to see how human beings could have made the practice so common. The vast majority of dairy cows, in the vast majority of situations, stand quietly to be milked, and it is obvious that they enjoy the experience. When a dairy cow misbehaves in the stanchion—as, for instance, when she kicks—there's a reason other than resentment of the experience, and it's usually easy to fix.

First, some reassurance: A cow's kick is not like a horse's; there's not the same power behind it, and it doesn't land as high. A horse's kick can be dangerous; a cow's almost never is. Still, it's not fun, and habitual kicking is incompatible with the kind of proximity you need for hand milking, so it's good to know why it's happening, and how to stop it.

If your cow kicks you, stop and give it some thought. The first thing to check is your fingernails. As hard as you're squeezing that teat, if your nails are long, you could be gouging her without knowing it. Multiple rings on the same or neighboring fingers, or rings with high-profile settings, may pinch or jab your cow; so might a flexible metal watch band. If the cow's teats are scratched or torn, start out the milking session gently; soon, a few good squeezes will have numbed the teat, and then she'll stop noticing. If it's fly season, she may be kicking at flies; try periodically flicking her belly and legs with your drying towel—it's draped over your knee or shoulder, remember—to help her combat this nuisance.

A cow may kick when she is startled, as when the barn cat runs under her manger or the dairy door is opened suddenly. You may startle her yourself. When a cow is in early training, it helps to prepare her for your hand on her udder by starting the touch at her back or thigh, running down her leg, and only then touching her udder. This kind of kicking is a nervous reaction and is seldom an ongoing problem. Sometimes—but rarely—she may kick in a fit of temper. If this is the case, she'll probably go on kicking. Whatever the circumstance, you don't want to let it become a habit.

There are several ways to prevent a cow from kicking. You can get someone to cock her tail—lift the base of her tail and hold it that way—which for some obscure reason will discourage kicking. If you put a rope on her near back foot and tie it to something behind her, so it can't come forward, you can cramp her reach so she can't kick you. If you put a rope around a cow's belly just in front of her udder and cinch it tight, the pressure on ligaments and nerves will usually stop her from kicking.

There are gadgets that you can buy that serve the same purpose. On the rare occasion that we have a cow offering to kick, we use kicking hobbles. This is a device consisting of two metal cuffs connected by a chain that can be tightened. The cuffs fit over the cow's hocks; with the chain drawn snugly across the front of her legs, neither foot can move far enough for kicking. The trouble then is that the cow's feet are rather closer together than you want for easy milking, but you can't have everything. You shouldn't need the hobbles any longer than it takes her to forget why she was kicking in the first place.

You can buy bigger gadgets, too, like kicking bars that clamp down on either side of her hips to discourage kicking. We've seen them used, but they struck us as awkward and clumsy, so we've never tried them ourselves.

Most often, a raised foot doesn't signify a kick so much as a fidget. You'll quickly learn to identify the shift of weight that signals your cow is getting ready to pick up a foot, so you're ready to fend it off with an extended elbow. It won't even interrupt milking. Truly, most dairy cows aren't inclined to kick; if you have one that is really determined, she disqualifies herself as a dairy cow and should be headed for the freezer.

Finishing Up

We have a couple more recommendations before you leave the dairy. We weigh the milk at each taking and record the results. We keep a simple spring scale hanging in the dairy for this purpose. Knowing how much milk your cow is producing over time gives you valuable information about your cow's health and state of fertility or gestation. Also, we tidy up the barn, especially if one of the cows has forgotten her manners and dropped anything we'd rather not have in there. Actually, we keep a 5-gallon bucket next to each stanchion; whenever a cow broadcasts, by lifting her tail or arching her back, that she intends to defecate or urinate, the milker grabs the bucket and holds it where it will do the most good. Material caught in this way will be dumped out in the garden, and the bucket returned to the dairy, thereby keeping the dairy clean and the garden fertilized at the same time. A clean dairy facilitates the production of sweet, clean milk.

Fitting Dairying into a Full Life

Mike and Gwen Sullivan,
Twin Cedars Farm, Jefferson County, Ohio

When Mike and Gwen Sullivan discovered grass-fed homestead dairy cows, it changed more than just what kind of milk their family was drinking. The blessings of intensively grazed dairy cows started the couple looking for ways they could expand their available acreage; and, when purchase was impossible, to move beyond conventional owned acres to an informal arrangement with the utility company. Their creativity and flexibility earns their family an abundance of fresh milk—"wholesome milk that we know is untainted" is a priority for Gwen—and multiple freezer steers every year. In addition, they have reclaimed twenty acres of abandoned land that, with its improved biodiversity, drought resistance, and productivity, is a blessing not only to their own neighborhood, but to everyone downstream. "Milking cows has been revolutionary for us," Gwen says.

Fitting milking chores into the schedule for a family with eleven talented and creative children, a family construction business, and, always, a large involvement in various mission works, was definitely a challenge. At first Gwen did virtually all the milking in their twice-a-day schedule. "Good thing Mike is a great cook!" she laughs. "Then I learned that I could milk only once a day without major consequences—I would lose some milk volume, but gain time in the afternoons."

Finding flexibility in the milk schedule taught Gwen how possible it is to match the milking schedule to family needs. "I found that we could change the milking time to fit whatever season we're in"—like midafternoon milking for midwinter, moving in spring to early mornings. For big interruptions to the schedule, Gwen makes sure she has nursing calves to take on the milking chore full-time while she's in the thick of things. Then, when life gets back to normal, the calves are put back on a twelve-hour-on, twelve-off schedule of milk sharing.

"Milking sets a rhythm for our family," Gwen observes. "I love to milk early and spend some time on the front porch praying and preparing for the day as I watch the sun rise. And when I wake the family, there's fresh milk and a healthy breakfast to start the day."

CHAPTER 7

Milk Handling

Milk is an animal product resulting from the metabolic process called *lactation*. When we consume cow's milk, we're drinking a bovine bodily excretion produced for the purpose of nourishing juvenile bovines. We get it by squeezing the back end of a cow. Any other explanation is whitewashing.

Are you comfortable with that?

You should be. As mammals ourselves, our very existence depends upon the milk our mothers produced. Milk from ruminants has been in the human food package since the first hunter who, having killed a young ruminant, smelled the rich, fermented milk curds in its stomach and was emboldened to taste. Viewed in this way, it becomes obvious that cheese has been a part of the human diet far longer than dairying; the voices of our ancestors overwhelmingly assert the fitness of bovine milk for human consumption.

In fact, it's obvious that milk must be pretty special: we go to great lengths to get it, even so far as to maintain large animals solely for the purpose of harvesting it. We've built entire food cultures around milk; we pay extravagantly for some dairy products. Milk ferments are naturally probiotic and an excellent source of the lactobacilli we associate with a healthy gut. It's not uncommon, though, to encounter dairy-handling protocols that give the impression that raw milk consumption is a form of Russian roulette, akin to drawing drinking water downstream from the city sewage system. Under the circumstances, it might be easy to get the impression that safe milk handling is necessarily a fraught and exacting science.

But it just ain't so. Oh, yes, our own early career as dairypeople was definitely colored by such ideas, enhanced by the fact that the milk of our freely foraging goats was flavored by every extreme and pungent phytochemical our very biodiverse Appalachian woods had to offer. Under the impression that these strong flavors were the result of careless handling, we became milk hygiene martinets, double straining and flash cooling in an effort to tame down our exotic dairy products—all without success. But while in one sense, these efforts were wasted—they didn't alter the flavor of our honest caprine-and-phytochemical dairy products one iota—their failure cured us of thinking delicious milk was primarily a matter of processing. We gave up over-finicking milk handling and just strained and cooled our clean, fresh milk, and the results taught us a lot: first, that if you don't like goats' milk, you shouldn't milk goats; second, that milk quality depends primarily on the health and diet of the lactating animal, and only secondarily on the milk-handling conditions.

Milk safety is one thing, and easy to achieve with a little basic care; the conditions for producing milk that is sweet and delicious are somewhat more exacting. While you don't need to be afraid that your raw milk, when handled properly, will breed pathogens, care in processing is desirable if you want the most delicious milk and dairy products with the longest shelf stability.

So let's talk about milk handling.

Clean Milk Starts in the Barn

After animal care and feeding, the most significant factor in the production of tasty, wholesome milk is the matter of handling. In the barn, empty buckets are placed upside down, to avoid inclusions from the hayloft and barn swallows; udder cleaning is a regular part of the routine; the first squirts of milk from each teat are directed into the strip cup. Straining, whether it happens in the barn or in the house, takes place in a timely manner.

Contamination: What Counts?

This is as good a place as any to discuss milk inclusions, pathogens, and the human ick-factor. Really, where do we get this squeamishness? Biological beings ourselves, living in a biological world, we are yet convinced that living organisms, especially the ones too small for us to see, are Out To Get Us, and that the most exacting measures are necessary to prevent our deaths from food poisoning, infection, or both.

Let's get over this right now. Go out into your barn on a sunny summer day and find a crack in the wall with sunlight streaming through it. Look at the dust motes sparkling there and think about what they consist of: dirt dust from the earthen floor, hay dust from the loft, feather dust from the swallows grooming over your head, dander from the mice building nests in the wall. Manure dust, of course, from every form of animal life that calls the barn home. And all this is floating around in the barn and landing on everything—on, and *in*, including in your milk bucket.

Now, before you start to turn green and curl up into a ball, stop, take a deep breath—and swallow. Down the hatch go all those varied dusts, drawn into your mouth with your breath—*and no harm done*. You're designed for this. Wherever you are, with every breath you take, you ingest environmental airbornes. Some are benign, some have the potential to be malignant; either way, your exposure to them plays a significant role in your immune health. When you ingest barn dust, you're receiving homeopathic doses of germs, allergens, and just plain dirt, that serve an immuno-therapeutic function in your health. That all these invisible bugs are also jumping into your milk shouldn't worry you at all; you're already taking them in with every breath.

Macrocontaminants

Of course, there are things that you *don't* want in your milk. Your cow's foot, for example: if she gets that in the bucket, it's over. The pigs will be grateful, so don't bother carrying that milk up to the house. Likewise, a large, identifiable lump of manure in the bucket means a good meal for the pigs, but we do mean *large* and *identifiable*—small, dry flakes of whatever was clinging to your cow's belly are going to fall into the bucket on a regular basis, and do not constitute a reason to ditch the milk. For good reason: remember, raw milk is naturally probiotic; as it exits the udder, it is inoculated with bacteria especially suited to milk fermentation, and their presence in the milk is a guarantee of the milk's ongoing healthfulness. There's nothing so inherently pathogenic in that flake of—whatever—that you need to be afraid of it contaminating the whole bucket of milk.

Other than a mucky foot or a glob of manure, what is there that, if it falls in your milk bucket, means you should reject the milk for human consumption? Oh, all kinds of things; but they're things you'll have no doubt about, like bottles of weed killer, dead rats, your neighbor's heart medicine. Seriously, it takes a lot to spoil a bucket of milk.

While milk handling doesn't require great speed, delay in straining and cooling means acceleration of the rate at which your milk will ferment. If you want sweet drinking milk, it's a good idea to schedule your chore list so milking is one of the last things you do before you return to the house and clean up.

At the House

A dairy strainer is a big funnel with a removeable screen or filter, which makes it easy to strain and jar your milk at the same time. For filter materials, you can choose how picky you want to be, using anything from throw-away microfiber dairy filters to a clean handkerchief. We don't like throw-aways, so in our dreams we use one of those coffee filter baskets made of fine wire microplated with gold; alas, they don't fit our big dairy funnel, so, to our shame, we use disposable filters. Muslin, scalded between uses, would probably work just as well, or better; it's probably where we'll go eventually.

Milk usually passes through its filter quickly, but sometimes not, and then you must wonder whether your cow has a low-grade mastitis infection. This is not something to get worked up about, and we just make note of the fact here; we go into it in more detail

Simple milk processing tools.

in chapter 12, Health, Unhealth, and End-of-Life Decisions.

Storage Containers

We have not been going to all these lengths to produce the best, most nutritious and delicious food known to mankind, only to contaminate it by storage in unfit containers, so we do not store milk in plastic, ever. Blech. Glass and stainless steel are our materials of choice: glass has the advantage that you can see through it to determine whether your milk and cream have separated; stainless

Skimming cream with a ladle.

steel, on the other hand, is durable. Both are inert; unlike plastic, they won't leach chemicals into your milk.

Qualities we desire in a milk storage container: We want something we can get our hand into, so we can really scrub it clean; we want something with a neck wide enough to admit our favorite 4-ounce (115 g) ladle, for skimming; we want something that isn't going to break easily; we want something we can fit in the refrigerator. In the beginning, glass containers let you see where the cream line is, as a guide when you are skimming; later, you won't need a guide. These days, our favorite containers are 5-liter (1.3-gallon) stainless steel milk cans that fit on the average refrigerator shelf; the booby prize goes to those nostalgia-inducing glass milk bottles that are impossible to clean properly.

Note that the size of your milk storage containers is going to affect how quickly the milk cools, and that, in turn, will largely determine how long the milk stays sweet. Smaller containers cool faster than big ones. Note also, however, that wide-mouth half-gallon mason jars—the most readily available half-gallons—are too narrow to admit a dipper.

Cooling fresh milk in shallow pans takes advantage of evaporation and greater surface area to lower temperatures passively. Cream rises to the surface and thickens, making it easy to lift off with a shallow bowl or paddle. This is how your great-grandma did it.

Cleaning Up

After your milk is strained, you want to wash up right away. Dairy equipment that is allowed to sit with milk residues on it develops calcium buildup called *milkstone*, accretions that can harbor bacteria and accelerate the rate at which your milk sours. To be honest, we've never seen milkstone, and we don't think you will, either; but it's a good idea to wash your utensils right away, in any case.

In the beginning, we used special dairy cleansers; eventually, we realized they didn't really apply to hands-and-buckets methods. They're expensive, caustic, require special disposal methods, and you don't need them. Whatever you use for hand-washing dishes will work just fine. Air drying is preferable for all milking equipment; dish towels, even paper towels, are vectors for bacteria. Our ancestors rinsed dairy buckets in cold water, and put a lot of store in drying milking equipment in the sun. Wooden utensils were scalded. As they made cheese we can't begin to imitate, the path of wisdom would very likely be to reinstitute their practices.

Temperature Considerations

Contrary to modern impression, refrigeration is *not* necessary for safe milk storage. Cooling milk has two primary purposes: to delay fermentation and to assist the raising of butterfat. Sour milk is not spoiled milk, and it's not "bad"; it's just sour. It's perfect for making soda- and powder-raised breads and pancakes; if you let it sit until it thickens, then hang it in cheesecloth, you'll have a lovely lactic cheese. So, whether you cool your milk, and how you cool it, will be best determined by what you intend to do with it, and what resources you have available. Refrigerators are only just so big, so you're really going to want some options.

Destinations

When milk is cooled to 40°F (4.5°C), fermentation slows to a crawl, resulting in dairy that stays sweet longer. Since we're bringing in fresh milk every day, longevity is not important, but North Americans like their milk cold, so we refrigerate drinking milk, or, in the winter, we put it on the back porch.

Warm milk straight from the cow is ready for making cheese.

If cheese making is *not* among the day's intended activities, milk not needed for drinking or cooking is set somewhere it can remain undisturbed, to assist the cream in rising. This, after twelve to twenty-four hours, can be skimmed off for butter making. Cooler temperatures assist the butterfat to rise, but, with cow's milk, it's going to rise in any case, so this milk need not be refrigerated; instead, it can go in the back of the garage, on the porch, in the basement, or wherever would be coolest. One lucky farmer we know has a spring house right in the cellar; spring water running constantly through a stone trough provides the perfect place to set even large cans of milk for quick cooling and fat separation.

The destinations and uses of milk on a homestead are almost limitless, too many to explore in anything but a whole book dedicated to the topic, something we are working on presently.

Trajectory: Homestead Dairy

*Jay Smith, unnamed state
correctional institution, Ohio*

Incarcerated since 2012, Jay has invested over ten years in study and preparation for the homesteading lifestyle he intends to pursue upon his release. Gardening in a 4- × 10-foot (1.2 × 3 m) in-ground bed, he's trialed many vegetable varieties, but he places more emphasis on the food independence possible from forage conversion. "I'm sold on the value of the grass-fed dairy cow," he says. "Even a 'yard cow' would be worthwhile; exchanging bought-in hay for wholesome dairy products and garden fertility is a good trade-off."

Jay has attempted to project in detail how the milk from a single cow might be utilized. "Milking twice daily doesn't seem as daunting as what to do with all that milk; whether that's 1 gallon a day or 3 gallons [3.8–11 L] doesn't seem to make much difference. Feeder pigs seem to be the answer. I can imagine setting aside two days' worth of milk for the table, cheese, butter, and yogurt, and dedicating the other five days to feeding pigs. If the homestead has grass to spare, baby dairy bulls could be bought in to utilize surplus. If the milking chore seemed overwhelming, calf sharing and milking once a day might also be a solution."

Jay knows the value of community, wherever you are. "Prison is a good place to find that out," he says. "You aren't free to leave, so you have to work out relationships with all kinds of people. Can a homestead be so different? Finding value in the rejected, damaged, and downright objectionable may be the single most important skill for the homesteader today."

With real estate prices continuing to rise, Jay foresees a time when land purchase may be impossible. "We could have to explore alternative means of land access, and borrowing or renting land looks like the most available option for people with limited incomes. The land most likely to be accessible is wasteland," he observes. "Substituting self-sufficiency for earning ability may be necessary. Producing clean, nutrient-dense food is a known good; producing to exchange farm products for cash is a less reliable good, for both the people and the land."

CHAPTER 8

Cow Nutrition

Yesterday, today, and tomorrow, foraging ruminants can tie us to our local ecologies and neighborhoods with the dual bonds of need and affection. Despite many generations of breeding selection focused on grain tolerance and a metabolic inclination toward milk production at the expense of body condition, cows are still fundamentally grass-conversion animals whose dietary needs are best met with pasture plants, and the homesteader armed with this knowledge is in a position to run her whole farm on grass—and milk.

But many homesteaders today are going to be working with cows bred for generations to fatten or make milk on grain. Genetically, they're grass eaters; epigenetically, they come from a long history of eating grain. How are they going to perform without it?

Our own experience is that *cows thrive on grass*. Isabel, our first cow, ate a lot of grain, but over time, we realized that feeding grain was a crutch to support higher milk production. With Isabel giving us up to 8 gallons (30 L) a day, we thought we could sacrifice some production. It took us two years to eliminate all grain from her diet; she went on to bless us with many thousands of gallons of milk over the rest of her lifetime.

Since then, we've transitioned a good many cows from grain supplementation to an all-grass diet—and we've bought and then sold a few cows that couldn't make the switch. Yes, cows are made to live and lactate on grass alone. And most mammals, under conditions of limited calories, will reduce their milk production; their bodies, faced with a conflict between lactation and survival, will choose survival. But a century and a half of human selection has been favoring dairy cows that will produce milk beyond the point of starvation, and the upshot is that some dairy cows will lactate themselves right out of existence.

That's because it takes a lot of calories to keep an 800-pound (365 kg) animal up and running, and when those calories come from grass, they're going to take up a lot of room. Concentrated feeds—grain—require less space, so modern dairy cows have smaller rumens than their grass-fed sisters. They don't necessarily have enough space in there for all the forage their functions require; because of this, not every dairy cow is a good candidate for an all-grass setting. The homesteader who is looking, first of all, for cows that keep good condition on grass alone, needs to remember that the price for natural and ecological balance is moderate production; and she needs to select her cow with that in mind.

Terminology

First, vocabulary.

Grass

This refers to whatever is growing out there: volunteer, default, non-woody ground covers. Grasslands are very diverse, consisting of many species of broadleaf as well as grasses, plus vines, small bushes, briars, and even tree seedlings; or, with planted pastures, anything from a simple grass/legume mix to the most complicated "pasture cocktail."

Grass-Fed

As we use the term, this means that grass is all the animal eats. This includes grazing while the grass is actively growing; it means grazing while the forages are dormant; and it may also mean eating dry, human-harvested forages (hay, and, sometimes, straw). Less commonly, it may also include eating fresh forages of many kinds, cut and delivered to the animal in the age-old practice called *soiling*.

Grazing

The grazing cow harvests her own forage. It follows that at the same time she is selecting which species to eat—as well as what parts, in what proportions, at what stage of maturity—she is also fertilizing and irrigating. *Holistic grazing*, or *rotational grazing*, is frequent, usually daily, movement onto fresh forage in small paddocks. *Bale grazing* is feeding hay spread on the ground for the purpose of adding waste hay and nutrients to the soil.

Supplementation

This refers to additional dietary calories, so it does *not* include noncaloric supplements like salts, minerals, or kelp. It does include things that wouldn't normally be in a pasture, including most things that come in a wrapper: grain and grain by-products, alfalfa pellets, fermented forages (silage, haylage, baleage), and processing wastes like bagasse (spent sugar cane) and sugar beet pulp. It includes all the things that come in tubs, like protein licks and molasses.

The term *supplementation* may also be used for offering hay during periods of inadequate forage availability. In this case it's the pastures that need to be supplemented (by forage brought from somewhere else, because the grass in the pasture is unavailable or inadequate), not the diet of the cow (with concentrated calories, as, for example, to induce a higher level of milk production). On our farm, cows are grazed only—they are pastured year-round. Sometimes, especially in late winter, we supplement tired pasture with discretionary baled hay.

Good Pasture

Not surprisingly, commercial agriculture has different standards from those of the homesteader, as is easy to see from commercial preferences in pasture composition. Commercial pasture is measured by its ability to grow exportable protein: the faster it will fatten cows or make milk, the better the pasture, period. Homestead goals are both broader, and more selective: A good pasture is one that supplies the cow with food and medicine; its bioactive soil takes up nutrients readily—that is, manure and urine are assimilated rapidly; and it releases nutrients in bioavailable forms through its diversity of healthy grasses and forbs. A good pasture is self-perpetuating—it does not need to be replanted. Its species include those that will be drought and flood resistant; they are reliably those that thrive under periodic grazing. By contrast, commercial pastures favor planted species, which must be regularly reintroduced; this, if you think about it, is why they are *planted* species in the first place. It is probably obvious by now that our own preference is for native, diverse, largely perennial pasture. Let's look at why.

Native Forages

A native pasture consists of plants that have been in the area for millennia (natives), or plants more recently introduced but that have shown over a long period that they can occupy a sustained place in the local plant communities (naturalized species). They're what's already growing out there, or what will begin growing if you initiate a holistic grazing program.

Native pastures in general share some other characteristics.

- They're generally a mix of perennials and annuals, emphasis on perennials.
- They're very diverse—that is, they include dozens to hundreds of species.
- Their diversity includes species with many characteristics: short, tall, and medium-height; those that prefer cool conditions and those that prefer warm; ditto wet/dry, long day/short day, heavy impact/light impact, and so on.
- Along with this diversity of species come other things, like a longer growing season, greater productivity, wider nutritional and dietary appeal, and so on.

It follows that, over the long haul, native pastures under holistic management will deliver more food and health with fewer inputs— *time and attention excepted.*

Holistic Grazing

Where that time and attention come in is at the level of pasture management. Good grazing for the benefit of your grass-fed dairy cow and the improvement of your (probably indifferent, at present) pastures will require time and attention. While holistic grazing may be said to be *rotational* (the cow is moved over the pasture in small paddocks one after another, that is, in rotation), the size, location, and composition of those paddocks is no machinelike clockwork. However, it does begin with very simple rules.

- Graze small paddocks, of short duration, with visible impact, followed by a long rest, with complete recovery before the next grazing.
- Remember that there is no bad rotational grazing—only good and better.
- Hold her tight and move her fast.

But these simple rules imply regular (even daily or twice-daily) visits to the pasture; regular (even daily or twice-daily) paddock moves; observation of the effects of grazing, and the rate and nature of forage recovery; observation, daily and ongoing, of the grazing animal; thoughtful analysis of the implications of what is observed, and the application of that analysis to future grazing decisions. The shorthand is: good grazing requires frequent moves, and the attention that goes with interest and commitment. If your life doesn't give you time for this kind of devotion, a cow probably isn't for you.

Grain Supplementation

People offer grain to cows for various reasons: to increase or maintain milk production; to increase or maintain cow body condition; because they are told they must; because they haven't thought about it; because it's fun. But it should be obvious that, since nature doesn't grain cows, grain isn't necessary. Although the overwhelming majority of commercial dairy cows are fed "concentrates," and even cottage dairying has an established history of periodic grain feeding to enhance milk production, the norm, worldwide and throughout history, has been to manage ruminants on forage alone. It is the conversion of cellulose, that ubiquitous yet elusive molecule, not of grain, that has made ruminants so valuable throughout the centuries.

And it is the privilege of the homesteader to rediscover those wonders of good dairying that were accomplished before the advent of cheap, subsidized, grain-based feeds. Increased milk yields notwithstanding, feeding supplements has many disadvantages—economic, metabolic, and holistic—and all of them play a role in our own choice against supplementation. For the sake of our farm's cash flow, for the sake of the cow's health, for milk quality, and for the sake of our land's ecological balance, we avoid feeding grain. Until we have found out what cows can do on grass alone, we won't really fathom cows; and until we understand cows—not modern, modified, concentrate-dependent cows, but hardy, resilient animals bred to express their natures—we won't know how best to partner with their gifts. For these reasons, we do not supplement with grain.

What About Treats?

So, what's the difference between supplementation and treats? Primarily, it is a matter of how much you are offering. Conventional cows are grained in proportion to their milk yield, at a rate of about 25 percent of their base caloric needs per gallon of milk produced. By that calculation, a cow producing 2–3 gallons (7.5–11 L) of milk per day needs about one-and-a-half times the calories as the same cow, not lactating. Supplied in the form of grain, this might be in the neighborhood of a gallon of grain per day for those 2–3 gallons of milk. In that case, a pint of grain—one-eighth of a gallon (about one-half of a liter)—over the same period looks like a treat, albeit a substantial one (about 1,700 calories).

But what about two pints? Or three? What about the temptation to offer a second scoop when the cow becomes impatient? How carefully are we measuring? Bearing in mind that many standard feed scoops are calculated to hold 5–6 *pounds* of grain (2.5–3 kg), you can see how easily treat-time could verge on substantial-supplementation-time. Supplementation sometimes has a place in dairying, but an unconsidered use of grain is sloppy and uneconomical, and it's not good for your cow. Obviously, it's important to know what you are doing, and why.

How Supplements Change Your Milk

While adding grain to your cow's diet may result in higher milk yields or a longer lactation, there are other, perhaps less desirable, effects of supplementation. The carbohydrates in grain are utilized differently from the long hydrocarbon chains of cellulose, and that difference is expressed in the composition of milk. Fat content comes from the digestion of long fibers (cellulosic material), so grass-produced milk is higher in butterfat than grain-produced milk. Grass milk has higher levels of beta-carotene and other vitamins and micronutrients. And our own experience tells us that grass milk is much sweeter than grain milk.

And while it's beyond the scope of this book to go into all the negative side-effects of grain feeding, note that the higher acidity of a grained cow's rumen means she can end up hosting acid-tolerant bacteria in her gastrointestinal tract, and some of these—like *E. coli* 0157:H7—are pathogenic to humans, and you don't want *them* hanging around the dairy.

At one time, it was more common to supplement dairy cows with roots and squashes than with grain. If, in the pursuit of ecological holism, you begin growing things like turnips, rutabagas, or pumpkins, for instance, with which to supplement your cows, keep in mind that strong smells and flavors transfer quickly from rumen to udder, and we mean *quickly*—like, in a matter of minutes. A bit of onion grass snatched from the fence line on the way to the barn can spoil a whole bucketful of milk, no kidding; so if you're offering pungent supplements, feed them directly *after* milking, so there's plenty of time for the volatile phytochemicals to be eliminated or metabolized before the cow's next trip to the parlor.

Also, barns where grain feeds are on offer are more attractive to disease vectors like rats and wild birds.

Feeding Grain: When and Why

Tune in at almost any time and you'll find us talking about the virtues of grass, grass feeding, and grass conversion into milk, meat, and manure. The blessed gift of the ruminant is her ability to turn cellulose into proteins, fats, and sugars.

Cows. Don't. Need. Grain.

Wait, though. We're not going to say it often, but we will say it: there *can* be a place for offering your cow caloric supplements—read: concentrates—read: grain feeds.

Yes, really. In a world in which most livestock are glutted on petroleum-derived corn and soybeans, it's usually more necessary to assure people that cows as a species *don't* need grain, never need grain, aren't made for eating grain, than to spend much time on the occasional utility of grain feeding. But it's true: there are times when feeding some grain may be a good choice, allowing us to strike a balance between the Natural Pattern ("cows don't need grain") with some specific human situation. Let's think of a few.

- Your family is establishing a working homestead in the mountains above Gallup, New Mexico. You have next to no grass at present, the only available hay being of questionable quality, shipped in from two hundred miles away and costing the moon. You're building soil with bale grazing and pastured pigs and poultry, and your family wants to produce as much as possible of its own food. Offering a pound of grain per gallon of milk produced increases your cow's caloric intake, and, given your so-so hay quality, lets you sleep at night. Yes, by golly, sounds like a little grain feeding is a good thing: more milk, more soil, and more grass—without turning your cow into a federally subsidized petro-derived grain conversion unit.
- Due to space constraints or other conditions, you can keep only one cow. You're buying local, native hay, made late because the farmer baled his higher-priced alfalfa–timothy meadows first. It's a cold winter, and you're trying to maintain your cow's winter production for your milk-loving family. She's bred for a fall calving, so you want her to lactate through the summer, and maybe a little grain will keep things going until spring green-up, when production would generally improve anyway. Go for it.
- You live in town, your homeowners association forbids live-stock, but you're so convinced of the benefits of the family cow that you've hidden one at the back of your lot, hoping that, if you're discreet, the neighbors won't turn you in. She's got a tenth-acre exercise lot surrounded by stockade fence, and she's hay-fed year-round—again, on native hay of less than the highest quality. Grain is what makes you feel good about keeping her under such close confinement; maybe, too, it helps her maintain her lactation.
- You've been searching for months, and the only cow you can get your hands on is grain-fed; do you buy her, or pass her up on the principle of all-grass or nothing? We'd buy

the grain-fed animal as a stop-gap measure, while continuing our search for an all-grass cow.

See? It's not as simple as Grass: good; Grain: bad. Situations vary. Cows as a class *don't* need grain, and too much grain is bad for them. It's also expensive, and produces inferior milk and flesh. But a little grain, offered judiciously, might give your farm an edge in a time of stress, and it might be what makes the difference between thinking *yes, we can keep a cow*, and *no, I'm too scared to try it.* We think keeping a dairy cow is a very good thing, and if the only way you can do it is by offering some grain, then offer the grain.

Mathematical Certainties

In the end, the only things you really know are the ones you experience yourself. Over the years, our own experience has taught us that in the husbandry of living things, as in art, it is better not to rely too much on numbers. Mathematical accuracy is possible, when it is possible, because numbers are abstractions; every two is identical to every other two. But living things, even within a species, are all different from one another, with the result that in nature, the more we know, the more we know we don't know.

Natural patterns are reliable, but the details vary. So, on the farm, few questions are ever completely settled, and we'll never be able to sit back and put things on autopilot. It is fortunate, then, given the depth of the average homesteader's unknowing, that living things like to go on living. They want to help you help them stay alive. And they've had a long time to work out good tactics for both individual and community survival.

Building Soil, Forage, and Family

Ben and Molly Macik, Project Regenerate
the Land, McGaffey, New Mexico

Ben and Molly Macik call their homestead in the mountains above Gallup, New Mexico, "Project Regenerate the Land." Ben grew up there and remembers when the land was green and the streams ran all summer. Then poor land management stripped away all but the most drought-tolerant groundcovers, turning the streams into gullies and leaving the soil almost bare beneath the Ponderosa pines.

Bringing animals back turned all that around. After two years of pasturing chickens and pigs began a slow transformation of their dry, rocky soil, the Maciks were fired up with enthusiasm for the potential of soil regeneration by means of animal pressure, and when they added a family cow to the mix, things really got moving. Encouraged by daily paddock moves, grass came in thickly behind their holistically grazed bovine. "The sheer abundance of manure and milk from just one cow is astounding," Molly says. "The milk cow was our revelation."

"This kind of regeneration is definitely not 'low-to no-inputs,'" she points out. "A milk cow doesn't like desert shrubs, but we like milk. Paying for hay to keep a milk cow alive is very expensive in our area, but it is so worth it to our family."

Living off-grid and far from town, Ben and Molly are used to keeping their focus close to home, so the regular milking chore isn't an obstacle for them. Molly milks with the baby on her back, while the older children, too small to be left alone in the house, play on the trampoline. "Granted, my life and schedule do revolve around the cow," Molly admits. "But

I don't find that to be a negative thing. I think it's a great way to live."

"I don't know that an easy life is something we're shooting for," she adds. "We want to have a good life; and a good life is fulfilling and challenging. Nothing is more fulfilling than watching the wind blow over the mountains while you milk your cow, say your prayers, and your baby reaches over your shoulder to pat your cow's belly, and says, 'cow, cow'! I wouldn't trade that for anything."

CHAPTER 9

Lactation

The morning you go out to the pasture and the cow that was still pregnant at bedtime is licking a wet, wobbly calf, life changes— radically. In ways you've never imagined, you're about to become bone of this land's bone, flesh of its flesh, a member of the biological community that extends from the subsoil to the stratosphere, and a partaker of its fate. Laying down soil organic matter, nurturing plant communities and the animals that depend on them, and taking your own sustenance from the abundance that results remakes you according to an essential human prototype that is as old as our species.

Holistic Protocol

Nothing illustrates more effectively the tensions between the goals and practices of commercial dairying and the God-given nature of the cow than the maelstrom of benefits, problems, and just plain issues presently associated with lactation. How much milk a cow should make, what she should be fed, and how to deal with the dozens of health issues that are endemic in commercial dairying are topics that could and do fill volumes. There are thousands of websites on dairying, from tiny farm blogs to USDA bulletins, each with its program for exactly how a lactation should go; and every program is different, even radically so.

With your new milk cow just freshened, you might want to have the different programs explained to you, with their goals, rationales, and results, so you can make an informed decision about best management practices. If so, you're on your own. Thirty years of keeping dairy animals has taught us that sometimes the only way to get a clear picture is to do your own research; sometimes doing your own research is how you learn not to expect a clear picture.

Our own dairy protocols are the result of doing the things that seemed to make sense in light of our strong conviction that Nature works, taking as our starting place the assumptions that, since cows are ruminants, their proper food is forage, and since they are mammals, they make milk. That isn't research, it's anecdote; but it's anecdote that is reinforced by many, many other anecdotes, all supporting those two assumptions about grass and milk. There are lots of other ways people keep dairy cows; these are ours. With this understanding, we share our experience with you.

Coming Soon

A cow's lactation begins even before she calves. Late in pregnancy—just how late varies—a dairy cow will begin to show signs of her coming parturition (calving) and lactation. *Bagging up* describes how a cow's udder enlarges—partly with milk, but more from edema—a few weeks or days before she goes into labor. This is usually most dramatic in a first-calf heifer; her bag will swell so much it looks shiny and she has to swing her legs out to walk around it. Older cows fill up noticeably as well. Over the same period, you'll observe that the cow's vulva is enlarging: the labia are thickening, and the whole orifice wobbles when she walks. You may also notice, in the last day or so before she calves, that her *pin bones drop*—a description of how it looks when her pelvic tendons soften and stretch, allowing the peaked bones on either side of her tail head to move out and down to facilitate the calf's exit. There are things you can be doing—or not doing—in these last weeks to help her have a successful calving and lactation.

Feeding Changes

Traditional wisdom has always said to reduce a cow's caloric intake in the last weeks before calving—not enough to starve her, but to establish a slightly negative energy balance. In recent years even the commercial establishment is falling in line with this idea. It turns out fat cows are more likely to have metabolic issues in the first few weeks after calving (see chapter 10, Calves and Calving). Lower-quality hay or pasture in the last few weeks of pregnancy will help keep those excess pounds off your cow; if you've been offering concentrates, they should be withdrawn at this time. For us, lower caloric intake happens automatically with spring or fall calvings, since during these periods pasture quality is naturally lower and cows are on the leaner side. It's worth saying, though, that we have calves arrive at other times of the year without seeing any problems, maybe because an all-grass diet is less prone to making cows too fat in any case.

Stanchion-Training: The Human Factor

In chapter 6, Milking, we looked at making your cow comfortable in the stanchion; now, as the time of her calving approaches, it would be good to spend some time training yourself to be comfortable with your cow. Hang out with her; put her in the stanchion and brush her. Sit next to her, with your legs under her, and handle her udder. Become familiar with her proximity, so that when something unexpected happens—she resists entering the stanchion, or she fidgets, or carries on in any way—you don't elevate the situation with your own case of nerves. Take the time to foster this relationship, so when you need to work with her intimately, you'll be known quantities to one another, friendly denizens of a safe, familiar world, and neither of you will panic.

The First Four Days

Calves can arrive at any time of day or night. We're not always present when a cow calves, so if we show up at milking time and there's a new calf in the pasture, it's guesswork what time the calf actually hit the ground. Knowing about when a calf arrived is significant because we don't milk mama for the first few hours after calving. She's got other things to think about, and other demands being made on her milk supply, and we can wait. But we don't want to wait too long: her comfort and her milk supply will both be affected by how soon and how often she is milked.

How much milk a cow is making when she calves, and how fast production ramps up after calving, is different for every cow, and for every lactation. In any case, for the first few hours or days, milk production isn't enormous, so there is less urgency when it comes to efficiency and thoroughness with those very early milkings. Increasing the demand on her milk production gradually gives your cow time to begin lactation with less metabolic stress, and our own protocols are directed toward this incremental onset. The first three milkings after calving help encourage a slow start.

Starting Off Gradually

We start milking about twelve hours after parturition. So, if we find a calf in the morning, mama's first milking will be that evening; if we find it during evening chores, we'll milk her in the morning. If we show up for chores and there's a new calf up, dry, and frisking around, we assume he's been there several hours and we just overlooked him, and we go ahead and milk his mama right away. Taking some milk out at this point helps ensure that, as the cow's production goes up, she doesn't end up impacted (full of milk, with nowhere for it to go), setting her up for mastitis (see chapter 12, Health, Unhealth, and End-of-Life Decisions).

If you are electing to separate the mother and her calf and bottle-feed the baby, you may still want to start by giving the calf twelve hours with mama. No one knows better than she does how to dry and stimulate him, launching him well.

Days One and Two

Whether you are planning a once-a-day milking routine, or intend to milk the traditional two times a day, for the first few weeks you should definitely milk twice a day. You don't yet know how much milk your cow is going to make, nor how much the calf, if he's with her, is going to take; so your hands, every twelve hours, are the safety valve that makes sure your cow doesn't get impacted.

First milking. For the first few days or weeks after parturition, your cow's udder is going to look and feel distended. This is partly milk, of course, but to a large degree it is simply edema. She's not about to burst, so this isn't a rescue operation trying to get every drop of milk out before her udder explodes. On the contrary, all you're trying to do with your first milking is take some pressure off. You don't want to take too much out, too fast; milk fever—the common name for calcium deficiency—results from a too sudden demand on a cow's blood calcium levels, and it's most common in early lactation. Milk fever is a serious condition that can be fatal, and beginning lactation a little more gradually helps avoid it (for more on milk fever see chapter 12, Health, Unhealth, and End-of-Life Decisions). To this end, at your first postpartum milking, don't squeeze until nothing more comes, but stop when the udder starts feeling less full.

Second milking. At the second milking, take a bit more than at the first milking. You're still just lowering the pressure; stop when the flow of milk has slackened considerably.

Third milking. Take it all! From this point on, you want to get every last drop she has in there. Milk production is a matter of supply and demand, so you want to make sure there's lots of demand.

Milking Technique

No one is a pro right from the start. History must be full of cows that were milked a little sketchily at first. Don't worry about it! You're not trying to induce this cow to make the absolute maximum volume possible (at least, you shouldn't be). If your first few weeks together include a little impaction and low-grade mastitis, you're getting a cheap education, and no harm done. Just go out there twice a day and squeeze that cow. Take your time. If she stops letting down and you've only taken a little milk, keep trying. Put a little bag balm on her teats and gently strip her (see page 90 in chapter 6, Milking), drawing two fingers down a teat, trying to get a few drops of milk. When those drops appear, keep going! Soon she'll "let down" again, and you can continue to milk normally.

Don't expect the calf to do all the work. This cow is a dairy animal, and she's the product of many generations of breeding for an animal that produces more milk than her calf needs. Even if your plan is to have a moderately productive cow on once-a-day milking, don't think you can start out by letting the calf do all the milking, and you'll take over when you feel like it. In early lactation her production will be rising, and with only the calf to take it, she'll be in trouble. At best, leaving it all to the calf will get you a lower-production mama; at worst, your cow will go through weeks or months of impaction and mastitis, and may even end up with permanent damage to one or more of her quarters.

Colostrum: First Milk

For the first three to five days of her lactation, your cow will be making *colostrum*, a specialized food for her calf that acts as a laxative, clearing his gut of meconium (black, sticky baby poos) and getting it ready for milk digestion. Colostrum is full of probiotics. It won't taste like milk, and it won't act like milk, and we don't much care for it, so what isn't consumed by the calf, we freeze for use as a tonic for sick animals—it's the best cure we know of for pneumonia in piglets. Some people drink it, or make it into pudding; we haven't been so inspired yet. A cow will only make colostrum for four or five days, so after the eighth milking (four days into the lactation), we consider what's being produced milk and keep it for the house.

You may wonder what happens if you're milking through a calving—that is, if you don't dry off your cow between lactations. Some folks worry that a cow that isn't dry between calves won't make colostrum. We can't speak for the whole species of bovines past, present, and future, but we have milked right through lots of calvings and all those cows produced colostrum, as we could tell in late pregnancy when suddenly we couldn't get cream to churn into butter for any consideration (one of the peculiarities of colostrum), or by the extremely firm curds we saw in our cheese making (another peculiarity). Colostrum can be identified in other ways, as well: it is thicker than ordinary milk, tends to be yellower, even orange, and its flavor is bitter.

Milk Production Ups and Downs

A commercial dairy cow's milk production curve describes a quick rise and then a long, gradual decline, held as high as possible for as long as possible by feeding lots of concentrates. This pattern does not describe the lactation of a grass-fed cow. Because her production is determined by what forages are growing, by your management decisions, by what the weather is throwing at her, by her state of gestation or parturition, and, if you are calf-sharing, by the persistence of her calf, a grass-fed cow's lactation will have ups and downs. Here are a few.

- As the seasons progress, the nutritional value of various forages shifts. Some things go up in carbs and proteins; other things go down. The water content of the forages varies with the seasons. Your cow's grazing selectivity makes a lot of difference to her milk production.
- As your calf matures, his milk consumption goes up—and then down. If you're not milking consistently, an older calf's reduced demand will result in an overall drop in production.
- Paddock decisions affect milk production. Experience will teach you how big to build paddocks in each season for maximum production. Likewise, paddock composition—the actual plant species and state of maturity—has a significant effect on production.
- Your cow's genetics will certainly play a role in determining the duration and productivity of a lactation; some cows will dry themselves off after nine months or so; some, especially if they are not bred back, can lactate for years.
- In spring, when the forage greens up again, a cow in mid-lactation may double or triple her milk production.

Nutrition, as you might expect, plays a big role in the volume and composition of your cow's milk. Some cows, as we've noted, are genetically inclined to very high milk production, and will produce milk even to the detriment of their own body condition. For the all-grass cow, with supplements off the table, loss of body condition may be addressed in a couple of ways. Offering larger paddocks lets a cow eat more, if she likes, and it lets her be more selective about which forages she chooses to eat. Dropping back on milk demand is the second option; you might take a cow nursing a large and demanding bull calf down to once-a-day milking for a while; or you might wean the calf. This goes doubly if you've got a grass cow acting as a nurse mother for more than one calf. If you're in doubt about what a too-thin cow looks like, check online for body condition score charts for dairy cows.

Seasonal variations in forage quality and quantity will have a big effect on milk production. Our own forage abundance follows a rough sine curve, with higher production in late spring and early summer, and again as temperatures moderate and rains return in the fall. Summer grass tends to be ligneous (woody), so while the volume is high, the nutrient value is less so. Stockpiled winter forage is of limited quantity—of course, since it's not actively growing—so the volume of milk produced is less, even while the nutrient value tends to be high (milk from winter stockpile may have almost twice the "components"—non-water constituents—of summer milk).

TAD or OAD?

How much milk a cow makes, or how much you would like her to make, are questions that will help you determine milking frequency. Remember, production is, in part, a function of demand; so twice-a-day (TAD) milking, all other things being equal, results in more milk than milking once a day (OAD). Early in a lactation, and when there is no calf nursing, TAD milking maximizes production, and, along with thorough milking technique, reduces the incidence of congestion-related mastitis. Later, though, it is possible to go down to OAD, either because less milk is desired (reduced demand will result in reduced supply) or because production is already low enough that milking OAD is sufficient accommodation for the milk being produced.

You may wonder, if incomplete milking and the resultant udder congestion sets a cow up for mastitis, how it is possible to go from TAD milking to OAD without causing an infection. We would avoid attempting it at the top of a cow's lactation, unless she was a very low producer or she was nursing a calf, but it's not a problem later on. Dropping from TAD to OAD is a once-and-you're-done event. You drop the desired milking time and, on the new schedule, recommence complete milking-out. You see? A half-day of congestion (the skipped milking time), then regular and thorough evacuation of the udder.

Late Lactation and Drying-Off

Then there's inducing the cessation of lactation, or *drying-off*, which is a different matter.

How long you milk your cow, and whether you dry her off before her next lactation, is mostly up to you—and her. Commercial dairy protocol calls for drying a cow off two months before her next calving, and there can be good reasons to do this, but they may not be applicable to your situation.

You'll hear it said that a cow "needs a rest" between lactations. Well, certainly, if she's been pushed with concentrates, meds, and hormones to produce very heavily for ten months, she may well need time to dedicate some calories to her new calf and her own health before she calves again. Our all-grass cows don't produce on that scale, though, as their good body condition in late gestation typically demonstrates, and often they can produce good calves and strong lactations without a dry period between freshenings.

Still, there may be reasons for giving your cow a dry period. If this is the case, you want to make her dry period at least eight weeks long. Why?

Cows make milk in response to demand. So, if you keep milking her, she'll keep making milk; if you stop milking her, she'll stop making milk. But mammary congestion (letting milk back up in the udder, as occurs with incomplete milking) creates the conditions for mastitis.

If you stop milking (the extreme case of incomplete milking), you create the conditions for an infection; in fact, you can be pretty sure your cow is going to end up with mastitis; and you really, *really* don't want your cow to freshen (calve) with mastitis, because if she does, the condition will get worse.

So, to avoid congestion-exacerbated mastitis is the second reason for giving a cow a two-month dry period. She needs to get over any infection *completely* before she calves and freshens again, since a cow that bags up with even a mild case of mastitis will, all too easily, become a cow with a serious case of mastitis. Even for your moderate-production all-grass cow, freshening with low-grade mastitis is likely to result in a serious udder infection. Lesson? If you're going to dry off your cow between lactations, do it with plenty of time for her to stop lactating completely and get over any mastitis before she bags up again—eight weeks, at least.

Why dry off at all? There can be any number of reasons. Maybe your family wants a rest from the daily routine of milking and dealing with milk; maybe you're planning a vacation away and don't have anyone to take over milking for you. Maybe, like the commercial dairy world, you don't want to bother with late-gestation colostrum. Maybe you have a high-production cow, or a low-conditioned cow, that you think needs a rest.

But then again, maybe you don't like the idea of going without milk for two months. Our first drying-off period was enough to convince us that we never wanted to go without milk for so long again. Next to being without unlimited, fresh drinking milk, we consider the daily chores a tiny inconvenience. If you feel the same way, maybe you'll decide not to dry off your cow; or maybe you'll get a second cow.

Going to the Next Level

Jennifer and Tobias Stephens,
Little Learning Farm, McDonough, Georgia

Jennifer and Tobias Stephens and their twins were already seasoned homesteaders when they bought their first Jersey dairy cow, Honey. Jennifer had found our work through a Homesteaders of America conference, where we presented information on how a dairy cow could truly nourish the whole farm. Jennifer remembers, "I couldn't take my mind off the idea of having a closed loop where our dairy cows would be providing much more than just milk. They would also be providing us with meat and fertilizer, and improving our low-quality land."

"Saying yes to the dairy cow was also saying yes to us selling our five acres in metro Atlanta for a home on twenty-five acres in a semirural area." But the new land was far from ideal; smothered in second-growth pine and mixed hardwoods knit together with muscadine, poison ivy, and greenbrier, it was almost impenetrable. The family got to work with the tools they had on hand: one chainsaw, and about a dozen goats. And, of course, their cow, Honey.

The first few days were not propitious. "We had never dealt with an animal her size. As soon as we got her off the trailer, Honey booted our son into the air with her head." He wasn't hurt, "but we knew that we had our work cut out for us." Reclaiming pasture and retaming Honey became the family's dual goals.

"She knew we were afraid of her, and for months she took advantage of our fears," the family remembers. "We realized we needed to let her know that we were in control." Conquering their nervousness, the family established a regular routine of milking and grazing. "Through time and repetition, we got through to Honey and earned her respect. Now she provides our family with fresh milk, butter, and cheese. She feeds our chickens, dogs, and garden at least once a week. Honey has been a wonderful addition to our homestead."

Having no outbuildings, the Stephens at first milked in the pasture. Recently, they constructed a small barn. Jennifer recalls, "For years, I milked my goats and cows in the open with just a head gate and a bucket to sit on. On rainy days, I would walk straight to the chicken yard to bless the poultry with the dirty milk. Nowadays, I can have clean milk no matter how bad the weather is. I'm also sheltered from the wind, rain, and sun. I am so grateful for the luxury!"

CHAPTER 10

Calves and Calving

Birthing is a prerequisite to dairying; a dairy cow isn't a dairy cow until she has a calf.

Birthing

Most bovine births are uncomplicated, and mostly we aren't there when they happen. Cows are good at having babies, and they don't generally ask for help or let us know what they've got in mind, so chances are we show up for chores and baby is already on his feet and looking for a meal. Sometimes the whole show is over before we get there; in that case, chances are the baby is already bedded down with a full tummy somewhere out of sight. If we can't see the calf, there will be other signs of what has been going on in our absence.

- Afterbirth—either on the ground, or hanging from the cow's vulva.
- Mama's transition from full-of-calf to suddenly empty is hard to miss, and for a couple of days her profile as viewed from the front will be startlingly deflated.

One clue that a calving has taken place may be a large, wet circle in the pasture where all the plants have been grazed extremely short; this is a giant "nest" where a tiny, wet calf making its first attempts to get to its feet won't get tangled up in tall forage.

Searching for a calf that has hidden itself away is a hopeless task; they can be completely invisible behind two crossed grass stems. We used to make ourselves crazy searching for newborns, convinced they might have wandered off. Eventually we learned that's not going to happen; they'll show up at mealtime. So don't worry if you can't see the little guy—yes, it's time to open presents and you want to see what this one looks like, but anticipation is half the pleasure, and by the time you've done the rest of the chores and put in a couple hours of real work, he'll be visible. A good general rule is "if mama is happy, everything's fine," but you do sometimes get a neurotic mama cow that works herself up about nothing.

Note: In case it's not obvious, there's no need for birthing to happen in a barn. The pasture is probably cleaner, and your cow will most likely prefer that setting. Weather is only an issue if it is *very* bad—monsoon, hurricane, or blizzard. Actually, we've had cows calve in a blizzard with no problem for mother or baby.

The vast majority of calvings happen without a hitch. In twenty-five years of keeping cows, we've only had to assist at two births—no, make that one; the first time, the cow would have been better off without us. There are occasions, though, when a little assistance is in order. If a cow is pushing for more than two hours without producing a calf, she probably needs help. And, yes, you might be able to figure out how to pull a stuck calf, or turn a malpresentation, but if you've never been doula to a cow before, we'd suggest that first time calling someone with more experience, like the vet. One new skill at a time.

The First Few Hours

Once baby is out, his first task is to fill his stomach. Watching a newborn calf struggle to get on his feet and find the teat can be a real trial of one's self-control, and the impulse to get involved may be powerful. Don't. It is remotely possible that the cow and calf really won't figure it out—but only remotely, and meanwhile your near presence, not to mention your fingers in the vicinity of his mouth, only serve to confuse the calf.

It may be an hour or so before the newborn nurses successfully. If you feel that things are taking too long, you could put both animals in your handy stock-panel corral, where there will be fewer distractions. Feel free to hang around for a while until you see the calf make sustained contact with his mama's udder; or, if you move on to other things, come back in a couple of hours and see how they are getting on. Mama should be calm, and the calf should have a full tummy (when he's standing, his sides are filled out, not hollow).

If a calving takes place in a paddock with electric fencing, it can be a good idea to turn the fence charger off for a few hours. A calf will not infrequently end up on the wrong side of the fence, which isn't really a problem, because he can slip right back in again, but there is a remote chance that he could get tangled up, and that's not only unpleasant, it's dangerous. Even though our fence chargers don't pack much of a punch (0.5 to 3.0 joules), a small animal that spends any time stuck in a fence getting shocked is going to be seriously compromised. You might reduce Mama's anxiety levels by enlarging the paddock to begin with, so her baby is less likely to slip out.

Calf Health

Here are a few other considerations in the first few days after birth.

Umbilical Care

In the commercial dairy world, it's usual to douse a calf's navel with iodine to prevent possible infection. Just spraying it or painting it on isn't enough; you're supposed to saturate the umbilical cord. It's not hard to do; you grab the calf, slip a cup half-full of iodine over his umbilicus and against his belly, then tip him up on his behind. This protocol makes sense when the birth-setting renders it likely the calf will be lying in mud or excrement, but our calves are born on clean pasture, and we never dip umbilical cords, and we never have umbilical issues. Maybe we're lucky, but it's a fact that beef calves born on pasture manage just fine without disinfectant.

Feeding

If you're wondering whether your calf is getting enough milk, first look at his sides. If his belly (the space between his last ribs and his hipbones) is full, he's fine. Another way is to check his poos. In the first day after birth, a calf will pass one or several stools that are dark green, brown, or almost black; this is meconium, the substance in his intestines while he is in utero. Subsequent stools should be yellow or orange, and fairly thick. Green stools may be an indication of dehydration, either because the calf is getting inadequate milk, or because he is scouring.

Scours

This is the term for diarrhea in animals; scours will soil his tail and back end and smell bad. Green, runny scours can be an indication of dehydration; a calf can get dehydrated pretty fast with scours, so this demands attention. You'll want to watch and make sure he's getting some good nursing time. White stools are also supposed to indicate scours, but sometimes a healthy calf can pass white stools that are solidish (and therefore not scours). We've never had a scouring problem with a mama-raised calf.

Retained Placenta

Failure to pass the afterbirth is not uncommon in cows. It's not the emergency it would be for a human being; in fact, cows can wander around for a couple of weeks with an incompletely passed afterbirth hanging down behind, without any negative health effects. It can come away in pieces, or even as a necrotic white discharge, and no harm to mama. If your cow hangs on to her placenta, don't pull on the externalized part; the weight of the externalized afterbirth helps draw out the part that hasn't detached yet. Oldtimers used to tie half a brick to the hanging placenta to add a little traction.

Stillbirth

Just a note here: sometimes, for no apparent reason, you'll have a calf that arrives dead, or that dies soon after birth. The beginner is apt to assume the death is due to some ignorance or oversight on her part, and blames herself. Maybe—but probably not. Cows calve unassisted all the time, mostly to calves that thrive; when they don't, the reason is usually (although not always) beyond our remediation. This is the moment when you find out how determined you really are about this homesteading thing. Attaching yourself to the wonderful life-giving forces of nature has its ups and downs, and you can't avoid the downs. The farm saying goes "if you're going to have livestock, you're going to have dead stock," which is another way of saying "stuff dies."

Calf-Sharing
and Bottle Babies

We've assumed your cow will nurse her own calf, but that isn't necessarily the case. Calves may also be bucket- or bottle-fed; there are advantages and disadvantages to each method.

Hand-raising dairy calves away from their mothers is standard practice, and it's how we started. When a calf was born, we took it from its mother and taught it to drink milk from a bottle (see "Bottle-Feeding," page 144). Not only did this let us reserve more milk for household use (since the calf couldn't help itself ad lib), it prevented the mother from deciding she preferred nursing her calf to letting down milk for the bucket. Although feeding a calf meant an extra chore, we were happy with the results—until the first time we let a cow raise her own calf. My goodness, what a difference! Our lazy, contented bottle-fed calves looked like sloths next to that rocket-fueled, mama-raised beast. We'd never seen such a level of health and well-being in a bottle-fed calf, and it made us determined to leave babies on their mothers whenever possible.

The following concepts will help you determine your own calf-raising protocols.

Calf-Sharing

This is when you let your cow raise her own calf while you are also milking her, and it's not just good for the calf; it can also be a worthwhile arrangement for people who want built-in milking help. Here's how it works.

A newborn calf is too small to be relied on to take any great volume of milk, so, to encourage milk production and keep his mother from getting impacted, in the beginning you'll need to milk twice a day.

The calf's appetite grows with him, and within a few weeks he may be nursing enough that your combined milkings, morning and evening, aren't yielding much. You might choose at this point to leave the pair together full-time and milk just once a day, but in that case, depending on when the calf chooses to nurse, sometimes you'll get a whole lot of milk; sometimes you'll get nothing.

For consistent milk for the homestead, you can also shut up the calf at night, and milk mama in the morning to get half (twelve hours' worth) of her production, leaving the calf with her the rest of the day to get his half. Don't worry about the twelve-hour gap in the baby's day; bottle calves are fed on a twelve-hour rotation with no issues.

Getting a Break

Once a calf has demonstrated that, given the chance, he'll take everything mama is making (that is, you experience two or more successive milking times when his mother comes in empty), you have the further option of sometimes letting him do the whole job. Say your daughter's on-farm wedding is in June, and your Devon calved in early March. By the middle of May, you have to separate mama and baby at night to get any milk for the house. If you're desperate for more time for wedding preparations, and you can stand the thought of not getting fresh, grass-rich milk every day, you can drop the milking chore entirely while you're busy with the wedding, and leave the calf with his mother full-time. He'll take up the slack while you're busy; then, when you're ready to resume your milking chores, you'll shut him up at night again and get your share.

Do keep in mind that handing your job to the calf, like everything in nature, involves a trade-off of goods. Leaving the milking chore to the calf works well, but naturally as he gets older he's going to take more and more of his calories from grass instead of milk; sooner or later, if you're not there to keep her production up, you may find when you come back to the job that mama's lactation is tapering off.

Delayed Let-Down

Calf-sharing has its risks. While most cows are happy to let down milk for the bucket, a cow with a calf will often hold back some milk—yes, they can do that!—and since the last milk out of the udder contains most of the cream, the majority of your butterfat gets saved for the calf instead of going into the bucket. Makes for a nice, fat calf, but your breakfast toast may be a little thin on butter until the calf is weaned.

Holding back can even take more extreme forms, with the cow refusing to let down for the human beings at all. Without a let-down reflex, neither hands nor milking machine can get milk out of an udder, so failure to get a let-down is a real problem. One solution, as old as the hills, is to let the calf suckle to start the let-down. The human milker can then milk alongside the suckling calf, or remove the calf and milk alone.

A more long-term solution is to retrain the cow's let-down to include human beings. This may be accomplished by some dedicated time and determined strip milking (see chapter 6, Milking), detaining the stubborn cow in the stanchion and revisiting her every quarter hour or so until your repeated stripping produces a few drops of milk. That's your sign! She has only just so much control over the reflex, and once you get the milk flowing, you're making progress. It may take a few episodes before you overcome her prejudices, but it's a small price to pay for a cow that's going to turn your grass into hundredweights of milk.

Bottle-Feeding

You can buy big two- or three-quart calf bottles at the farm store; you may have to enlarge the opening in the nipple to get a reasonable flow.

A bottle-fed calf should be positioned with his neck down, his head back, to prevent aspiration of milk.

There's a technique to bottle-feeding a calf: the first few times you'll
have to straddle the calf's neck, clamp your knees behind his head, and
tilt his chin up. Stick your finger over the little air hole in the nipple
collar before you flip the bottle, so milk doesn't go everywhere, and
force the nipple into the calf's mouth—you'll probably have to pry it
open first.

So far, it's all pretty intuitive, but here's the detail that matters—
keep the calf's head down. You want his neck inclined downward at the
shoulder, then arched back a little to elevate his mouth (picture the
angle at which he would have to approach his mother's udder, slung low
between her legs). Neck *down*, head tipped *up*. The object is to avoid
any angle that would mean that milk not swallowed will run down the
calf's trachea. Aspirated milk can incline a calf to pneumonia, a leading
cause of death in bottle calves.

Milk, Formula, and Creep Feed

What will you put in the bottle? We'd say milk, but most bottle calves
are getting *calf milk replacer*, another term for soy-and-corn-syrup formula. It comes powdered in bags at the farm store, and costs plenty, but
not as much as you'd make selling the milk the calf would drink, which
is why people use it. Let us urge you to resist this temptation; formula,
even organic formula, is going to lead to a host of other problems, big
and small, and is no good substitute for real milk.

Bottle calves, even bottle calves getting good cow's milk, are prone
to scours; if this happens, we shake a raw egg into each bottle before
feeding. With advanced cases of scours, calves can become dehydrated
and weak. Electrolytes for bovines can be found in packets at the farm
store; so can bottles for *drenching* (force-feeding fluids). Since calves
with scours are often acidotic, we add a couple of tablespoons of bak-
ing soda to each quart of fluids in a calf drench.

Commercial dairies offer a creep feed of medicated grain to baby
calves starting in the first week. Milk subsidies are higher than the
cost of (also subsidized) grain, and commercial dairy cows need
grain-friendly rumens, so this makes sense for their setting. The feed
is medicated in the expectation that the calf will be dealing with infection. If you want to raise this calf for eating grass, though, milk is the
best food for him. First, it will establish the rumen biome he needs for
grass digestion; second, grain-feeding results in a small rumen, and a

grass-fed animal needs a large rumen for his bulky diet; third, do you want meds as a regular part of your calf's diet?

Separating the Cow and Calf

Whatever you decide about your calf-feeding strategy, sooner or later you'll need a way to separate the calf from his mama. A simple, inexpensive calf hutch can be made with a single 16-foot (5 m) welded hog panel. Bent into a circle and closed with carabiners, this is the easiest DIY calf pen going; it lets cow and calf see and smell one another, but there's no way for the calf to get his head out and nurse.

For a bigger calf, a 5-foot × 8-foot (1.5 × 2.5 m) pen is plenty big enough for overnighting, so long as the calf can be out during the day; for a long-term separation, such as for weaning, a larger pen is indicated, like the temporary corral discussed in chapter 4, Acquiring a Cow. (*Weaning* is any program for changing how or whether the calf is getting milk; hence, we may say *wean* for a calf's transition from nursing on mama to taking milk from a bucket, or for removing milk from his diet altogether.)

Whatever you use, it has to be very sturdy; the calf isn't going to approve of your plan, and he can get through, over, or under some pretty surprising barriers. If he gets out and can't find mama, he'll go looking for her; and if he doesn't find her, he'll keep going. It takes only one or two times chasing black calves through dense woods at two in the morning to convince you of the value of a strong weaning pen. For backup, we put a bell and collar on every calf when it comes off its mama; bells are cheap insurance.

When Milk Comes First

Keep in mind: the most important thing about a dairy cow is that she turns grass into milk and makes it available to human beings. For her to do that, we need to breed her and calve her out; we *don't* need the calf. Calves are darling, they are necessary for the propagation of the species, and they make great beef, but for homestead dairying, they are—pardon us—expendable. That expendability is why humans eat veal. Where there's a surplus of both grass and milk, raising the calf makes perfect sense, but if you don't have the grass or the milk to spare, find a new home for the calf. There are folks out there who buy dairy calves—we used to do it ourselves—and if it isn't practicable for you to raise this calf, don't feel bad about letting someone else do it.

Regenerating Farm, Family, and Neighborhood

Daniel and Carlee Russell,
Double Roots Farm, southeast Wyoming

All-grass dairying holds tremendous potential for regeneration. Dan and Carlee began with some pretty unpromising acres, but soon discovered the healing power of holistic grazing.

"Our seven-point-five acres had been a junkyard, a hound dog rescue, and an overgrazed horse pasture," Carlee relates. "Our alkaline, high prairie land tops out at 4,100 feet (1,250 m), with temperatures ranging from −30°F (−34°C) to well above 100°F (38°C). Fortunately, we discovered rotational grazing and the regenerative power of the dairy cow."

Today, in addition to nursing her own Angus-cross calf, their Guernsey cow Clementine provides 2 gallons (7.5 L) of milk daily, to feed eight people, five cats, and thirty-five chickens. "Everyone shares the workload," says Carlee. "Dad milks; the kids make yogurt, labneh, and butter; and I am learning to make rennet cheeses." Between all the fresh dairy, eggs, and meat the farm provides, they have cut their large family's grocery bill by over $10,000 a year.

As important as is her role in the Russells' food bill, Clementine contributes even more to their lifestyle and health. "Five of our children are adopted, and three suffered with severe digestive and skin issues when they came to us," Carlee shares. "Raw milk healed their skin and tummies—no more eczema, meds, or creams. In consequence of our total commitment to providing the family with fresh, raw milk, our college-age son has dubbed us 'lacto-vores' and our eldest daughter now homesteads off-grid." Their long-term plans? "A multi-generational homestead with three generations of family, farming together."

Benefits of this magnitude have a way of spreading. A neighboring farmer, at first dismissive of their work, was converted by their lush, green pastures. "Last summer, he built himself a chicken tractor," says Carlee with a laugh. "Next, we expect to see a dairy cow."

CHAPTER 11

Breeding

So many chapters on managing a grass-fed, homestead dairy cow, and we're only just now getting to a vital issue without which none of the other things matter: how do we get our cow bred?

Detecting Heats

The first step to getting a cow bred is to identify when she's in heat, or *estrus*.

First, a few salient facts.

- A cow's *estrous cycle* is roughly twenty-one days long, meaning she's open for breeding once in three weeks.
- *Heat* (estrus) is that period of her cycle during which she will ovulate and be receptive to breeding.
- The only reliable sign that your cow is in heat and ready to be inseminated (by bull or artificial insemination) is when she is in *standing heat*, which means she stands to be mounted. To be clear: when mounted by another bovine, she doesn't try to scoot out from under.
- A cow *ovulates* about thirty hours after standing heat is observed. Artificial insemination (AI) should take place

about eight to twelve hours after the initiation of standing
heat, but if your cow is sharing pasture with a bull, you can
leave things to him.

A cow that is approaching ovulation desires the services of a bull,
and she broadcasts that need in a number of ways. The first is one
we can't see: pheromones. Other bovines can smell when a cow is
about to be receptive to breeding. They'll sniff her tail, sniff or taste
her urine, and attempt to mount her. Any bovine may attempt to
mount a cow in heat: adult females, bred or open (not carrying a
calf); steers (castrated males); heifers (young females); bulls; or what
have you. The cow approaching estrus may offer to mount any other
animal, even a nonbovine. Does this promiscuity surprise you? It
shouldn't. Pheromones can be detected over long distances; if these
airborne hormones were his only clue, a bull receiving scent signals
from a large herd might have hundreds of tails to sniff in order to
locate the source, a big job even if all the cows agreed to line up and
keep still. Promiscuous mounting simplifies things by broadcasting
the location of the cow in heat with enough accuracy to narrow the
search range considerably.

A complete estrous cycle extends over about forty-eight hours.
There's a period of buildup to standing heat, followed by about sixteen
hours of receptive (standing) behavior, then twelve or more hours of
wind-down during which estrous symptoms are still to be observed.
Behavior during the pre- and post-receptive periods is pretty similar.

A cow moving into estrus may exhibit decreased milk production,
restlessness, fence pacing, or extra mooing. She may show minor
aggression toward other bovines—head-butting, for example—or she
may be coquettish, sidling up to other cows head-to-tail and presenting
her tail to be sniffed. In anticipation of copulation, she'll produce a lot
of clear, stringy mucus of egg-white consistency, which you may notice
sticking to her tail or rump, or pushed out of her vagina during a bowel
movement to pool on top of the cowpie. Her vulva may be swollen,
but if you haven't made a point of observing her back end recently,
you might not notice. Her belly may be smudged down one or both
sides by the front feet of cows that have tried to mount her; the hair
at the base of her tail may be ruffled from the mount and dismount.
You might catch her resting her chin on the spine of another cow, or

mounting other cows. If she has no other way of expressing herself, she may try mounting nonbovines, like the farm dog—or you.

During the period of actual receptivity to breeding, the cow will voluntarily stand to be mounted. In the absence of standing to be mounted by another bovine, mere mounting behavior isn't a reliable indicator of anything; it *may* mean a cow is nearing estrus, but it might just as well be an expression of dominance, or even mere friskiness. No, we're not kidding: when a cow mounts another cow, either, both, or neither may be approaching heat.

If a cow's estrous signs can be so confusing, how can you be expected to catch and take advantage of her hormonal cycles? Although individual signs may be ambiguous, daily, ongoing observation of your cow will acquaint you with her normal patterns of behavior. As she moves through her fertility cycle, you'll recognize variations, the same way you know when one of your children is sick, or when your spouse has something on his mind. During the breeding season, it will be most helpful if you build some time into your day for cow observation. There are worse things than an obligatory twenty minutes lounging in the pasture. Take a cup of coffee; pull up a comfortable stump and stay awhile. Mornings are best for catching a heat.

Once you've noticed signs of impending estrus, determining standing heat is pretty simple if you have more than one cow—the cows will show you. Lacking another bovine, though, you can check for standing heat yourself by leaning heavily on your cow's tail head, or bracing your hands on her hind end and giving a sudden shove downward. If, without being restrained, she stands to allow this, she's most likely in standing heat.

If you're milking your cow by hand and moving her fence at least once a day, you've got the most important tool for determining estrus: time with the cow. Don't hurry over moving fence; use your powers of observation while you're doing chores. Take your brain off screen saver and think about the cow as you work with her; you'll be surprised by the things you'll start to notice.

Paint and Stickers

Tail paint and heat-detection stickers are aids to help you determine that a cow is standing even if you're not there to see her being

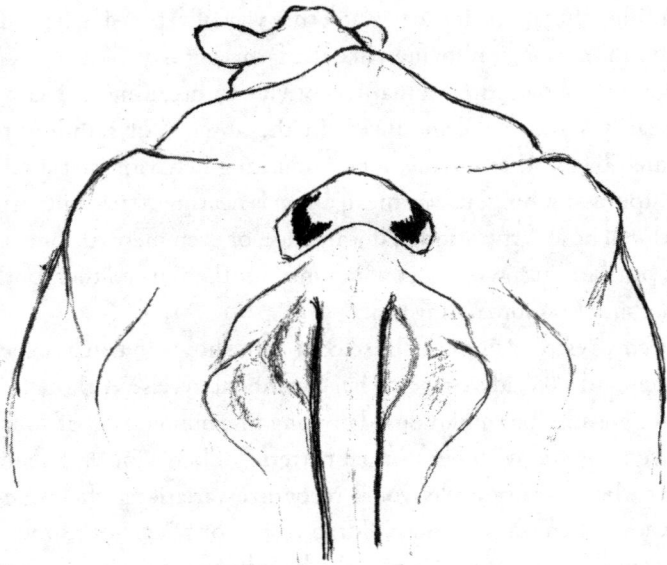

A "full swipe"—a sticker completely scratched from top to bottom—indicates that this cow stood to be mounted.

mounted. Paint is, well, paint, and you apply a good daub of it to the base of the cow's tail; heat-detection stickers are like scratch-off lottery tickets. If a cow stands to be fully mounted, paint or sticker gets a good swipe as the top cow slides off, and if your mind is on your business you should notice the next time you look at her that there is less paint on her tail, or the lottery ticket is heavily scratched. Find paint or stickers in a farm catalogue or online and follow instructions. We use heat-detection stickers to catch heats in cows we want to artificially inseminate, or to establish a due date for cows that are bred by the bull.

Although a cow generally begins to cycle within six weeks of calving, you may not find it easy to spot a heat right away. There can be a number of reasons for a delayed or "silent" heat. Weather plays a factor; hot weather seems to discourage shenanigans, and if cold weather keeps you inside, you may miss the signs. Heat stickers or tail paint can be very helpful under these circumstances; even if a sticker-timed breeding doesn't "take," knowledge of when your cow was standing tells you when to begin watching for the next heat.

Stickers and tail paint need to be renewed every few days; summer breeding coincides with fly season, and a constantly flicking tail will wear down a heat-detection device almost as completely as a mounting pasture mate.

Bull or Technician?

Bulls have been breeding cows for so long because they're really good at it. AI, no matter how skillfully performed, is not to be compared with insemination by a bull; so let's look at bull breeding first.

Bulls are not only better at detecting heats than we are, they're better at getting a cow with calf. They deliver literally hundreds of times more sperm than is contained in a semen straw, and they also have a longer breeding window after heat is detected.

But do you want to keep a bull on your farm? Things to consider: A bull in a small herd may have only one working day a year, but he eats every day, so if your pasture is small, there may not be room for him. Conversely, pasture needs to be pruned anyway, so if you have the grass to spare, his appetite is an asset.

If the primary goal of breeding is lactation, and the calf is destined for the freezer, the range of acceptable bulls widens, and your neighbor's beef bull might be in the running. Sharing a bull is not uncommon; when one or more neighbors keep a bull together, his impact and cost of upkeep get spread around. Ask how large a calf a potential bull generally throws, though. Small calves have better outcomes, but there are still folks who get a kick out of breeding for 150-pound (60 kg) calves, and you don't want to use their bulls.

Bulls as a class are no more dangerous than many other animals, but dairy bulls specifically have a deservedly deadly reputation. Statistically, Jersey bulls are the most dangerous animals in North America. On the other hand, we've kept Dexter bulls for ten years, and they've impressed us with their good manners.

AI

Artificial insemination used to be touted as the best thing that ever happened in livestock breeding, and there is something to be said for how it increases a farmer's available genetics beyond that of the local stock. The result in the long run, though, should not surprise people familiar with, say, ice cream. How many local creameries have gone bust since a little one in Vermont went global? How many creameries-in-potential never got past the dream stage? Sure, now you can buy Ben and Jerry's anywhere in the world, but the options for local ice cream have been reduced correspondingly. AI, while it makes "the best" bulls available everywhere, has resulted in a severe diminishment of the total available genetics, and permanent loss of diversity only takes a generation. A failure to value local adaptation becomes, too quickly, the complete loss of that adaptation.

We're not (necessarily) advocating for throwing out AI. Semen-collection technology is increasingly available to small breeders, including farms specializing in heritage breeds. For homesteaders landlocked by commercial ag, AI may be the best chance of accessing traditional livestock traits, and a little genetic material judiciously introduced doesn't have to mean total replacement of local with exotic.

Most commercial dairies use AI today, often with "sexed" semen—semen that has been sorted according to sex of potential offspring. This lets them breed their highest-producing cows for replacement heifers, using dairy semen, while inseminating the rest of their animals for beef-cross bull calves. The results are about 90 percent reliable for sex. Sounds like the best of both worlds, but anecdote, including our own, suggests that along with sex determination, sexed semen is delivering some funky neurological and health problems, and we don't use sexed semen when we use AI.

The question of whether to bull breed or use AI will be moot if you can't find a technician to do the job for you, and artificial inseminators don't generally hang out billboards. Ask your veterinarian if she can recommend someone, or reach out to local homesteader groups and online forums. Many commercial dairies keep their own AI tech on staff, but some use a roving tech whose name they might be willing to share. Some national semen supply houses have regional representatives who moonlight. If you don't find someone right away, keep looking— you're probably not the only person in your county who needs a cow artificially inseminated.

Supposing you decide that AI is the best choice for your cow this year, and you've even found an AI tech—you've still got decisions to make. Semen-supply catalogues have encyclopedic listings describing the many qualities of their studs, and your tech is going to ask you what you want. The list of qualities we seek is fairly short.

Grass genetics. It's funny to list this at all, because by definition *all* cows have "grass genetics," but in a stud book, it should mean that this bull individually, and his family line in general, have received minimal or no grain. Since we're going to feed no grain at all, avoiding genetic lines for heavy grain tolerance makes sense. But so far as we know, there's no standardized definition for *grass genetics*, so in the end we're just taking their word for it.

Long teat length. This metric doesn't apply to the bull, obviously! Instead, the reference is to his near female blood relations. We select for long teats; although teats can be too long for convenience, for decades dairy selection has been for *very* short teats, and we want to keep away from those. Four fingers long (as measured from knuckle to knuckle—about 3 inches, or 8 cm, or a bit longer) is about ideal for us.

High components. As applied to milk, *components* means "whatever isn't water": primarily proteins, fats, sugars, and minerals. Jersey and Guernsey milk (to name just two) are high in components as compared to Holstein milk; in other words, they are more nutrient-dense. Like teat length, this trait is determined from the bull's female relations. We select bulls graded for "high components."

Polled. In a stud book *polled* means naturally hornless, which is how we prefer our cows.

A2 homozygous. Although we don't presently have any strong opinions about A2 beta-casein, we do have strong opinions about heritage genetics. These tend to be A2 homozygous (A2-A2), so we also select for this trait.

Calving ease. Bulls that tend to throw small offspring will be described as such, often by some term like *calving ease*. Cows birthing small calves have fewer problems and need less intervention, so we consider this a desirable trait.

You'll be happier if you settle how, by whom, and with what to have your cow bred, before the time is upon you. Which leads to the question, "How do I decide *when* to have my cow bred?"

Cow Reproduction: Timing

Once you have good, home-produced, fresh raw milk on tap, you don't want it to stop. So, one of the first questions the novice dairy cow owner asks is, "How long will she keep making milk?" which, when you think about it, is just another way of asking, "When does she need to be bred to calve again?"

Beef cows are usually bred to calve in the spring, in order to take full advantage of the grazing season for nursing calves. But breeding a dairy cow is more about milk than it is about the calf, and your calving plans will reflect this.

How long a cow's lactation lasts is a function of many variables. These include, but are not restricted to, her breed, genetics, epigenetics (induced variations in her genetic expression), age, weight, forage type, caloric intake, frequency of milking, grazing patterns, whether she's bred or not, whether she's nursing a calf, and so on. Many cows of dairy breeds will lactate for two or more years after a single calving; some will not. The safest course is usually to plan for your cow to freshen every year.

Calving time can be calculated from the breeding date by the formula: "subtract three months and add a week." Breeds vary, with larger breeds (and larger individual cows) generally gestating longer than smaller breeds (and cows) by a few days, but by this simple reckoning we can determine that if we see our Jersey cow covered by the bull on August 9, we should expect her to drop a calf not too far from the middle of May.

You're going to want to be looking at next year's calendar before you send for the bull or the AI tech, so you don't plan your cow's calving for the same week as your daughter's wedding. We time AI according to considerations like weather—we'd rather see a cow calve in April than January—and forage availability—cows calving in March (or September) will have green grass for making milk. The bull times the rest, and generally, but not exclusively, gives us spring calvings.

AI dates, obviously, are a known factor. Mark your calendar. If you see the bull mounting a standing cow, that date also lets you calculate a due date. But you're not done yet. AI, in general, is about 75 percent effective on the first breeding, and bulls don't always score the first time. That means that as many as one in four cows bred will fail to settle with a calf. So, three weeks later you need to be watching to see whether your cow goes into heat again. We try to remember to put a fresh heat sticker on her eighteen days after breeding (by bull or AI), in case we're looking the other way at the significant moment.

Pregnancy Testing

If a cow shows signs of returning heat, it's likely that she failed to settle to breeding. If there's a bull in the pasture, he'll make it his business to try again; if you're using AI, you'll be calling the tech again. But nature loves variation, and sometimes a cow exhibiting hormonal behavior has actually conceived, and is just settling into her new pregnancy. How are we to know the difference?

We wish our grandfathers were here; they could see whether a cow was bred just by looking at her hair patterns. Until we figure out how to do that trick ourselves, though, we check for pregnancy either by blood test or urine dipstick. Both of these are cheap and easy options. It's easy to draw a blood sample from the tail of your cow while she's in the stanchion (watch a video). If you prefer the dipstick, you don't have to follow your cow around until she pees; you can induce her to urinate by stroking her escutcheon (the velvety skin just below her vulva). And the vet can do pregnancy checks, either by palpation (feeling the uterus through the rectal wall) or ultrasound, but it will be more expensive—much. In any case, if you want to know for sure that your cow is bred, wait the appropriate interval after breeding (it varies with the test), and do a check.

Sometimes, even when you've gotten a positive test, a bred cow fails to drop a calf. Test results can be inaccurate; cows can miscarry. Stress may cause a cow to *slip* a calf (the common term for miscarriage in bovines). We had a traumatic dehorning once, and the subsequent stress caused the cow to slip her five-month gestational calf. Sometimes a fetus will be genetically deformed or otherwise nonviable (like the "bulldog" calves of chondrodysplasic Dexters), and these will be slipped early as well. In a big pasture you might not see the aborted fetus; in a holistic grazing operation, it'll probably be obvious. On a small farm, it's always worth going to see what the vultures have found. Failure to calve is a big disappointment, and a serious hitch in the farm food and feed cycle; it usually qualifies a cow for freezer camp.

DIY Dairy

Roy and Pamela Foster, "the Gribble Place,"
Ontario, Canada

For Roy and Pamela Foster, jumping into dairying was a lot like taking a winter dive into nearby Lake Huron—bracing. Research and study had convinced Roy and Pamela that adding dairy animals would be the first, best step they could take toward feeding themselves and reclaiming their overgrown, abandoned northern Ontario farmstead. Arriving in early autumn, the animals were greeted enthusiastically by their new people.

"But winter was coming," Roy remembers. Hauling water by hand, mucking stalls, and carrying hay to their charges morning and night grounded, but did not abate, the Fosters' enthusiasm; still, it wasn't hard to understand why previous generations of farmers, loggers, and homesteaders had packed it in and gone elsewhere.

When the time arrived to breed their future dairy cow, the Fosters discovered to their dismay that the local AI technologist was going out of business. Once more, they reached into their fund of independence. "Our bull was still quite small, but we'd heard the saying 'where there's a hill, there's a way,'" says Roy. He built a ramp of packed manure for the vertically challenged bull and engaged their cow Queenie's cooperation. Although negotiations were a little awkward, his efforts were blessed with success, and the following April, Queenie presented the Fosters with a heifer calf.

Within hours of her birth, however, Roy and Pamela knew they were faced with another problem: Queenie's lethargy and cold, cold ears told them she had milk fever. Without a veterinarian or source for supplies, the Fosters had to think fast. "We drove several hours to find the calcium and IV flex we needed," they remember. With the help of a video, they taught themselves to administer a bovine IV. By morning Queenie was on her feet, chewing her cud, and nursing her calf.

Dairying in Canada includes some real challenges, but the Fosters are certain it's worth the effort. "We like milking cows," says Pamela. "Grass farming sinks our roots deep in the community."

CHAPTER 12

Health, Unhealth, and End-of-Life Decisions

Sir Albert Howard, considered by many to be the founder of the modern organic farming movement, famously said, "The whole problem of health, in soil, plant, animal, and man, is one great subject." This being so, every living thing is a perfectly unique product of its always-in-flux environment—which precludes the possibility that anyone, whether scientist, doctor, farmer, or agronomist—will have a tidy formula for bovine health. Certainly, *we* don't.

We have always believed that health—our own, our animals', our farm's—is Nature's default position. We reject the idea that well-being waits upon some scientific breakthrough, exotic miracle food, or novel health regime. In the pursuit of an inputs-free, ecosystemic farm, we've tried to provide our animals with settings and food that are natural to them, believing that the result should be health—a long shot, maybe, given that most animals, farms, and environments today are somewhere on a spectrum from damaged to devastated, but it's worked. Our soil, pastures, gardens, woods, livestock, and family enjoy a high level of health and productivity. If we ever doubt that our holistic principles are playing out as we hope, all we have to do is look at a farm that's doing things the other way.

Still, in this world there are failures of health; there is injury; there is death. Those are the facts, and if we're going to keep animals, we'd better get comfortable with them; there will be days when our willingness to continue homesteading depends upon it. When those days come on your farm, we hope the following pages will be some help in sorting out what is happening, whether it's something you should do something about, and what that something might be.

Appreciating the Veterinarian

Isabel, our first dairy cow, was with us for twelve years. She was sweet-tempered, pretty, and gave us thousands of gallons of milk. She was also, as we gradually came to realize, an unmitigated constitutional disaster—no exaggeration. Of eight pregnancies, she delivered only two live calves. She had six "down" episodes, each lasting from a day and a half to more than a week and requiring some kind of round-the-clock response from us. She was a grain-fed, barn-kept, over-indulged princess, and we learned a whole ton from her. We never want a cow like that again.

Isabel was also our introduction to the local large-animal vet, who thought homesteaders were a menace to any and all livestock unfortunate enough to land in their way. As our acquaintance with her progressed, our weird, indefensible ideas about forage-fed, low-production family dairy cows only made her more skeptical. It took years for us to come to a good understanding, years in which we still needed a vet, and we were lucky that Polly's commitment to the local farming community and to food animal well-being meant she didn't strike us off her list of clients. We continue to learn from our friendship and professional relationship with her, even as our foundational principles persist in diverging.

Like us, you're going to need a vet. Sooner or later, if you keep livestock, something will happen that is, or appears at the moment to be, beyond your skills, something that may mean life or death to your cow, and you'll want the opinion of a professional. Once you get that opinion, you'll need to know when you disagree with it, and why, and how to do your own thing without mortally offending her. Maybe you have a similar relationship with your doctor; you're glad she's there, but you don't always follow her advice.

Reach out before you need anything. Vets won't love you for assuming that, just because you're in the neighborhood, they have no option but to accept you in their practice. They're expected to do things that are messy, uncomfortable, undignified, and dangerous, so they like a little consideration; a uterine prolapse on a rainy night in March is even less attractive when the cow owner is a total stranger. Be proactive; arrange to meet first under more congenial circumstances—not, maybe, where you'll appear to be asking for advice you're going to ignore ("hi, meet my all-grass dairy cow, what do you think?"), but for something routine, like a pregnancy check. Let her see that your farm has a plan (even if it's one she won't think much of); show consideration for her comfort and convenience; nod your head a lot. Bring her on the farm once a year even if you don't need anything, just to keep the relationship going; a single vet visit is a cheap retainer fee.

And don't lose sight of your principles. Your vet's education is probably oriented to a different kind of farming, one that uses different methods to accomplish different goals. You may be doing things she doesn't think are possible, to accomplish goals she can't relate to. Pursuant of those goals, you may need her skills, even if you won't always follow her advice.

You can do a lot of your own veterinary work, and sooner or later you'll probably need to. For example, it doesn't take a veterinary degree to learn to run an IV on a "down" cow. We have been lucky that our vet expected and encouraged us to acquire some DIY skills and helped us develop them; vets are busy, and some things can't wait.

What Does Health Look Like?

From the beginning it was hard—no, impossible—to find examples of the kind of farming we wanted to do. Most dairy cows are on a regime of hormones, medications, and supplements. Exceptions are treated as norms (hence, the standard is to vaccinate whole populations for rare conditions), and the only acceptable direction for production to go is *up*. If we insisted on believing that the default position of Nature is health, we would be on our own when it came to defining what "health" looks like. Just for a baseline, here are some of the things we've learned.

- If a cow runs and kicks up her heels, she's probably healthy.
- If her ears are up, her eyes are bright, and she's grazing and chewing her cud, she's probably healthy.
- If, rising after a good rest, she arches her back and takes a nice stretch, she's probably healthy.
- If her coat is clean and shiny, slick and oily in summer, fluffy and sparkly in winter, she's in good health.

In periods of doubt or discouragement, keep these signs in mind.

The Best Defense

There are two ways to approach illness: one is, you got sick because you contracted a bug; the other, you got sick because your resistance was low. There's validity to both approaches. Native American populations exposed to the novel European smallpox virus died in droves; the common cold lurks in our communities unnoticed, periodically breaking out to express itself in some folks as a sniffle, in some as pneumonia. The majority of cow problems are going to fall into the second category—illness following stress.

Just a Boost

Healthy cows are like healthy people: they mostly stay that way. For the small things, cows have a remarkable gift for healing, and usually problems go away without diagnosis or treatment. When cows pick up a germ or receive a minor injury, they generally heal on their own. When something seems to be dragging them down, though, they benefit from a little consideration as much as we do, so if it's really cold out, maybe a couple of nights in the barn are indicated; if a cow is limping, it's helpful to shorten the distance to food and water. Immune-response boosters like raw apple cider vinegar and garlic in her drinking water are a good first line of defense when a cow is puny. They, and other simple remedies, will usually be all you need to deal with problems.

Pasture Wisdom

Cows on native pasture learn to identify and use nonfood plants for their health benefits, which is the same as saying cows are good botanists and herbalists, which they are. If you manage pastures holistically, imitating the graze-and-rest patterns of a healthy feral herd, the cows, instead of concentrating on high-energy forages, also sample the therapeutic and medicinal species that support natural health. They learn to identify and employ those species, and they pass that knowledge on to herd mates. The longer you keep a cow—or, better, a family line of cows—in the same place, the more skillfully they employ pasture wisdom. As a result, a great many of the endemic health issues seen in conventionally grazed and confinement herds just aren't going to bother a holistically grazed cow.

An Incomplete List of Cow Issues: The Little Stuff

Here are a few common conditions you can probably manage without medication or vet visits.

Mastitis

Mastitis is an inflammation or infection of the udder, usually expressing as clumps or strings of matter on the screen of your strip cup. Less acute signs are slow-filtering or salty-tasting milk. Most often, a single quarter is affected without involving the rest of the udder. Mastitis is much more common in cows that are grain-fed, machine-milked, and/ or high production; not surprisingly, it's endemic in commercial dairies and relatively uncommon in homestead cows. In the latter case, it tends to clear up in a few days without assistance. Tradition associates mastitis with lack of gentleness or thoroughness in milking; completely emptying the udder at milking time is supposed to speed recovery. The overwhelming evidence past and present is that mastitic milk is not a health hazard for humans consuming the milk, but if you don't want it on the table, the pigs/chickens/dogs/cats will be glad to get it.

Most cases of mastitis are minor: you don't know why they came, and you don't know how they go away. Occasionally, you'll get a case that lingers for weeks or months, or just flares up periodically. The infection is local, not systemic, so you don't really have a sick cow, but something is preventing her immune system from throwing it off. Raw apple cider vinegar and garlic may help boost her resistance; herbal and homeopathic remedies you might use for minor malaise in humans should also be helpful. Our own experience has overwhelmingly been that mild cases go away on their own.

Flocculation on the strip cup screen.

Sometimes, though, you'll see more serious cases. A cow being dried off between calvings is going to develop impaction mastitis when her human stops milking her. With a couple of months between lactations, she should have time to recover before the next freshening, but if she's dried off late, she may still be harboring bacteria when she bags up again, and the intramammary pressure, plus pathogens, can lead to trouble. One or more quarters may be hard and hot. Instead of milk, it may produce something thin and clear, with strings of pus or mucus, or a thick paste-like substance that's hard to squeeze through the teat. Your vet can do a culture to determine if there's an appropriate antibiotic for the infection; there are lots of nonprescription and herbal remedies out there, too. We, personally, have found massage and frequent stripping to be the most helpful therapies in such cases, and have never intervened with antibiotics, but that's not to say that medication would never be indicated.

Most cases of mastitis will clear up in a short time, but some linger. A bad case can leave permanent scarring that blocks the teat canal, so you can't get milk out of that quarter, which is a catastrophe or a mild annoyance, depending on your attitude. A cow with a nonfunctioning quarter has a lower market value, but her condition doesn't necessarily result in a decline in production; in the case of an all-grass cow, especially, the other quarters can take up the slack. It's not something we lose sleep over.

In a quarter-century of keeping dairy cows, we've seen many cases of mastitis. Almost all were brief and had no long-term effects; a few resulted in nonfunctional quarters. None have compromised the well-being of the cow. It's not that something bad can't happen, it's just that it's rare.

Colds

You'll sometimes see a cow with a runny nose or a cough that goes away in a few days and doesn't require any action on your part. Rarely, though, a cough may last a lot longer. We had a cow that coughed occasionally when being driven up to the barn for milking. The vet examined her and found nothing amiss, but she coughed like that for about a year, then outgrew it and lived for many years after. If a cow is otherwise healthy—eating, chewing her cud, stretching when she rises, normal poos, normal temperature—we don't worry about a cold.

Feet Problems

Trimming a cow's hoofs can be a wrestling match, with the humans trying to get a rope on the foot in question and the cow surging about objecting, and we do it as seldom as possible—like, practically never. Hoof growth rates are a response to the kind of surface the cow is walking on: if the surface is soft, the hoof grows slowly; if it's hard, the hoof grows faster. This can be a good thing if the cow is doing a lot of walking, but if she's not, her hoofs grow faster than she wears them down and they end up long and curled. For this reason, confinement cows often come with messed-up feet; we avoid them and their overgrown hooves, crossed toes, and chronic fungal infections. Our cows, pastured year-round, seldom walk on pavement and have pretty feet that get them where they need to go. In a quarter-century, we've trimmed only a couple of times, and we're pretty sure even those were unnecessary.

Rarely, we'll see a cow with a limp, and if she's got a long walk back to the dairy, or a hoof-pocked frozen lane to navigate, we'll rearrange fences for a few days to make her job easier. These things have always healed up quickly, and we've never needed an intervention.

Cuts

Cows have an astonishing healing capacity, even recovering without issues from trauma as severe as a perforated gastrointestinal tract or a compound fracture. We bear this in mind when we're faced with cow injuries.

We've seen some pretty spectacular cuts over the years, mostly due to the ubiquity of corrugated sheet metal in farm-building construction. Hit it at the wrong angle, and it cuts like a knife. Our vet educated us: if a wound is gaping and going to pick up a lot of dirt, you want it closed, and stitches might be the best way; otherwise, it'll heal on its own. One of our cows brushed up against the duck house roof and put a 12-inch (30 cm) gash across her ribs, through dermis and epidermis. It gaped like a briefcase, but it was clean, and the vet left it open; it healed without a scar. A few years later, the same cow went on a spree over a strip-mined wasteland and had a dustup with some abandoned equipment, coming home with a 9-inch (23 cm) gash clear into the hip socket, damaging the joint capsule. Out of our pay grade, we thought, and called the vet, who said it would heal all right the way it was—and it did.

If we feel like we simply must do something for smaller owies, and things that are likely to get dirty, we squirt them with peroxide and apply liberal amounts of topical antibiotic salve (human grade, both from the dollar store), and they heal very quickly.

Abscesses

Sometimes a cow will get a localized infection at a puncture point and an abscess will form (it looks like half a golf ball inserted under the skin). At first we had the vet lance them; then we lanced them ourselves; then we just ignored them. Seems like if they need draining, they drain themselves, and if they don't, there's no point in making a hole in the cow. Haven't seen one in years; don't know why.

Internal Parasites

When we kept bought-in cows and orphan dairy calves, we occasionally saw evidence of intestinal parasites: things like low body condition; a rough, dry coat; and manure buildup at the base of the tail. Sometimes the vet cultured manure samples and gave us prescription anthelmintics (dewormers); sometimes we bought whatever antiparasiticals were at the farm store. Then we quit, and with years of holistic management, our soils have become more bioactive, our pastures more diverse, and our cows tougher and smarter. When the herd begins to look scruffy, we graze them along the tree line so they can access more medicinal plant species, and they slick right up; for decades that's been our only parasite strategy.

Keep in mind the difference between a regular practice and a rescue operation. Our animal health program consists of good pasture in holistic rotation, and our herd is made up of animals that do well in that program; parasite resistance is built in, and we very seldom use anthelmintics. If a calf doesn't fit that standard, we'll be finding that out her first winter, and she will go in the freezer the next fall; under no circumstances will she be bred. But before we'd let an animal die of its own unfitness, we'd worm it—once.

External Parasites

When we kept goats in the barn over the winter, we saw goat lice, but we've never seen lice on a cow. We can't remember seeing a healthy cow with any kind of skin parasite, not even dog ticks or deer ticks—and we spend a lot of time in close proximity to our cows. As with a suspected internal parasite load, if a cow is looking scruffy, we rearrange paddocks so she is getting some woods'-edge forages.

Fly strike is a maggot infestation that can happen when something—manure buildup on tail or rump, a wound, or constantly wet skin (as perhaps in a "down" animal—see pages 172–173)—attracts egg-laying flies. Hatching by the thousands, the larvae begin to feed off the flesh of the infested animal. It may be enough to use a local remedy like diatomaceous earth, alcohol, or mineral oil to kill the maggots, but in more widespread cases a topical antiparasitical may be called for. We, personally, would reach for the ivermectin.

Ringworm

This fungal skin infection strikes herds in our neighborhood every few years, and we are told it is carried from farm to farm by the local deer. We see it maybe one year out of three. We suspect a mineral deficiency is responsible for the animals' susceptibility; maybe one day our soil will be so bioactive that ringworm will be gone for good. Meanwhile, if the affected animal is where we can easily handle it, we might paint the spot with mustard; if the animal is too big to hold still and a long way from the barn, we ignore it. In both cases, it goes away.

Warts

Warts are viral skin lesions, and the story is they travel, like ringworm, with the deer. They look ugly but don't hurt the animal, and the cow's immune system will eventually deal with them. You can hasten the immune response by crushing a wart, like with a pair of pliers, which we never do (ouch). Once she's had warts, the cow develops an immunity, which is why they are most often seen on calves.

More Cow Issues: The Serious Stuff

Veterinary problems on the homestead tend to fall into three categories: minor things you deal with yourself; big things you call the vet for; and really big things that happen when the vet is out of the county and you have to figure it out on the fly with the help of YouTube videos and old issues of *Small Farm Journal*, *ACRES*, or *Stockman Grassfarmer*. The following fall into the third category.

Bloat

Grass digestion is fermentation, and fermentation produces gas, and gas takes up room. Ruminants would pop like balloons if they weren't designed so the gas can get out. Most of it is eructated (burped) along with undigested fiber ("cud") returned to the mouth for further chewing, so cows don't rupture.

Sometimes, though, there's an obstacle to the eructation process. The gases get backed up and the cow starts to inflate. That's bloat. Something has to happen to deflate the cow pretty quickly, or the internal pressure will increase until her lungs can't expand and she'll suffocate.

Positional bloat is pretty easy to fix. This happens when a cow gets "cast"—tipped over on her side and unable to right herself. In hill country, like here in northern Appalachia, a recumbent cow (resting on her sternum) can end up on her side, feet uphill; and with her weight downhill she's unable to get up again. Sometimes breeding shenanigans will end up with a cow on her side. With her head lower than her rumen, ruminal gases can't escape through her esophagus, and she begins to inflate. The fix is to push her back up on her chest, so the rising gases have a way to get out; or, if her feet are uphill of her body, pull her head around until her body is on the uphill side. This kind of bloat is situational, and if you're there, you can fix it; if you're

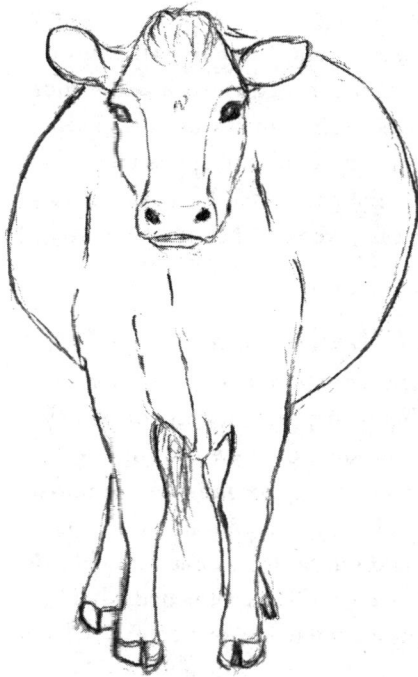

Left flank distended beyond the hook bone indicates bloat.

not, you'll probably lose the animal. Don't worry, though; it doesn't happen very often.

Feed-related bloat is just what it sounds like. If you're not offering grain, you're a lot less likely to see this kind of bloat, but it can happen, especially on frosted pasture or pasture with a whole lot of legumes like clover or alfalfa. If you're standing in front of your cow and her left flank (your right as you face her head) is distended to the point where you can't see her hook bone, she's bloating. What can you do about it? First, look to see if she's chewing her cud. If she is, then she's still able to burp, and her state is equivalent to yours after a large pizza: uncomfortable, but not in any danger—at least, not yet. If she's not chewing her cud and she keeps lying down and getting up, you need to get involved. Walk her; this will encourage the gas to move. Keep a close eye on her. If the swelling gets worse, you'll need to take action. This could mean running a stomach tube to evacuate the gas, or administering an oil drench to reduce the surface tension of her stomach contents, or, in

extreme cases, punching a hole in the cow's side with a pointed tube
(a trocar). In a quarter century we've seen only one case of feed-related
bloat (from a patch of frosted clover), and it didn't require intervention. If it ever does, we'll probably use a stomach tube and mineral oil
to reduce the foam and relieve the pressure (we'll watch a video first);
if that doesn't do the trick, we'll nerve ourselves up to use a screw-in
trocar. In either case, we would earmark that cow for an early trip to
freezer camp.

Two Kinds of "Down" Cow

Milk fever is the common name for *hypocalcemia*—calcium deficiency—and when it happens, it's an emergency. The most common
victims are high-production cows that have recently calved. If you're
lucky, you bring the cow up for milking and notice she's shimmying as
she walks; more often, you go out and find your lovely girl unable to
get up, or even laid out on her side looking like death—which is why
she's called a *down* cow. What's happened is her blood calcium level has
dropped to the point where her muscles can't function, and if you don't
do something pretty fast she'll die of heart failure.

If you catch it, and you have an IV flex and some calcium gluconate,
you can do a jugular IV on your cow and get her back on her feet, usually in a few minutes. It's nicer if you can watch someone else do it first,
which is why you want to be on good terms with your veterinarian, but
when the time comes the vet might be busy with someone else's down
cow, so it's a good idea to keep four bottles of calcium gluconate, a veterinary IV flex, and some large gauge (14–16 ga.) hypodermic needles
where you can find them in a hurry. Try the farm store, or order from
an ag supplier online. For lack of a real person to show you the first
time, a video tutorial takes three minutes, and it will be good enough;
there's nothing like a life-or-death emergency as motivation for doing
new things.

Over the years, we've dealt with more cases of milk fever than we like
to remember, and we'd just as soon avoid it—seems like it's always cold,
wet, and dark when we have to IV a cow. Our Dexter-cross cows don't
seem milk fever–prone the way the Jerseys were, but we administer a
tube of oral CMPK (calcium, magnesium, potassium, phosphorus)
or CalMagCo (calcium/magnesium/cobalt) paste within the first
twenty-four hours after every calving, just in case. These pastes come in

tubes, like caulking compound; you put a lead rope on the cow, cinch her tight to a post and get someone to hold her head up while you pry her mouth open, shove the tube to the back of her mouth, and push the stuff in. You want to do it a bit at a time, like maybe three goes, so she has time to swallow it. She'll hold on to the last dose and spit it out if you let her, so keep her head back for an extra minute. Don't give oral calcium to a cow that's "down," in case she's too far gone to swallow.

And if all this sounds too, too dire, bear in mind that we've never lost a cow to milk fever.

In early spring, when the grass is full of moisture—"washy," meaning it goes through the cow too quickly—you may sometimes see magnesium deficiency, or "staggers." In twenty-five years, we've seen two cases—wobbly cows that looked like they were drunk. Not pleasant. We made sure they didn't tip over and get "cast" (see "Positional Bloat," page 170), and gave them a tube of CMPK, and they recovered just fine. Every spring after that, we've made sure to have some high-magnesium mineral in the mineral pan, just in case.

Tests and Immunizations

Disease is real; human beings have come up with some wonderful ways of dealing with sickness, and we are grateful for them. We use them when we believe they are indicated; otherwise, not.

Here is our disease policy to date: we don't immunize cows. Yes, we know there are infectious and potentially fatal diseases out there. The dairy cow population in the United States carries a tidy load of endemic diseases like Johne's and bovine leukosis, and if you feel concern about them, we strongly suggest doing your own research. Ours indicates that while the vast majority of dairy cows go through life exposed to and even carrying these diseases, only a tiny fraction will actually get sick. Stress is usually the disease trigger, and our cows don't see a lot of stress.

So, we don't immunize cows. Neither do we test for disease. If an animal seems healthy, we assume she is healthy. To our knowledge, we've never lost a cow to a disease; we've certainly never seen an epidemic. Sir Albert Howard famously listed the characteristics of a natural community, and we recommend memorizing them all, but the last is this: "Both plants and animals are left to protect themselves against disease." Joel Salatin points out that natural immunity is what's left after a disease ravages a population (our words, not his); viewed in the short run, that might mean some dead animals, but a couple of generations later, it's eradication of the disease.

Cows Are Mysterious

The partnership between bovines and human beings is ancient, fruitful, and, in some cultures, even sacred—a thought that leads us naturally to the subject of cows and mystery. Cows under species-appropriate conditions—holistic rotations, all native/naturalized pasture, no grain—enjoy a great deal of natural health. Lots of things you might expect to bother them—like a necrotic placental retention—don't bother them at all. They are very resistant to infection; they heal quickly and completely after trauma. We don't really have to get very involved in the minutiae of their health.

Maybe that's why sometimes we have no idea what's going on with them; and when we say *we*, we mean ranchers, farmers, veterinarians, homesteaders, all of us. That marvelously balanced, just-keeps-going, give-her-some-grass-and-let'r-rip bovine will sometimes just. lie. down. And then not get up. And you won't know why. You'll call the vet, and she'll come and do an examination. She'll probably administer some calcium, magnesium, B vitamins, and dextrose; she may hoist the cow up with a winch or a tractor and try to get her to stand. And sometimes this will work, and sometimes it won't—but, either way, no one will be sure what was/is wrong, or why. And if the cow won't get up, you'll have to decide how long to let this go on. You could give her physical therapy and room service. It could go on for weeks. She might get up—we've certainly seen that happen, and more than once or twice—and she might not. If you slaughter her in the first couple of days, she'll make decent burger; if you wait much longer, her muscles will be atrophying, and she won't.

If this ever happens to you—and it may never happen to you—you should think about a few things.

Life is short. Cows are mysterious. Stuff dies. And it's OK.

Since this cow came on your homestead, she's been an immeasurable asset, mowing, fertilizing, producing gallons of the best milk, probably calves, hence, beef; she's been an education, a workout, a reason to be outside every day; you've learned to make butter, yogurt, sour cream, lactic cheeses; and you're a different person, knowing where food comes from, aware of your place in the universe, all because you and this cow have been taking care of one another.

Your relationship has been a phenomenal success. If it's time to say goodbye to this cow, know that you have both been blessed. Know that nothing lives forever. Know that the life you gave her was better than almost anything she might have had elsewhere, far better than a life of confinement, concrete floors, bushels of grain, medications, wormers, and machine milking. Know that you can offer her no greater dignity in death than that her body go on to grace your family's table and power your labors. Give thanks for all these blessings. And then go straight out and find another dairy cow to partner with; because the kind of person you are now, needs a cow.

Cows and People and Blessings

Chances are, you and your family cow are going to be working together for a good, long time. Before breeding programs turned dairy cows into gaunt milk-production divas, it was common to see a cow making calves and milk right into her twenties, even thirties. Not so common today; but it's reasonable to hope that a healthy cow may live productively right into her teens.

Still, like us, someday she'll start showing signs of her age. The deep sockets on either side of her tail head, the pendulous udder, the slow, dignified tread that never, anymore, breaks into a gallop or a frisk—all speak of the accumulating years. Spring comes and she drops a calf, as ever; her production may be almost as good as when she was five or six; but summer gives way to autumn, and you begin wondering how your girl is going to manage the winter.

Have you any idea how a cow dies, if no one's there to help her along? Mostly slowly. When you weigh 800 pounds (365 kg) or more and your legs are just sticks, the first sign you're losing the battle with time is that you can't get up. There will be scars in the turf around you to show you've tried—and tried. Your head's up and your eyes are bright, but all the grass within reach will be eaten down to dirt and you'll be thirsty when someone brings a bucket of water, because you can't get up to find it. And if no one intervenes, you'll stop trying to get up, and soon the circulation in your legs will be inadequate, and gangrene will set in. Or you'll keep trying to get up, and you'll tip over and bloat. Or the coyotes will find you. It's not pretty.

Or the story could be different, because there are people involved. We can see age happening, see the day coming when that dicky hip will play her false, and we can make choices about what her final hours will be like. It's possible for there to be no break between sun and grass and company, and lights out. *We* may feel bad, but she won't. Then we'll harvest her with love and respect, and hope someday ourselves to be buried where we can push up daisies so her great-great-great-great-granddaughters can eat them.

In the dairy cow and human cooperative, it's a pretty moot question who is working for whom. If someone put our meals in front of us every day, gave us rubdowns, looked after the budget, and made sure we were comfortable when the weather wasn't, it would have to be a creative argument that convinced us we were slaves. To us, that sounds more like family.

The truth is, cows and humans have helped one another for eons. Bovines access cellulose for human use; humans, looking ahead, provide the guiding hand that fends off overstocking and famine. Together, we can form the landscape into shapes ever more fruitful, abundant, and beautiful, for all the creatures—not just people and cows—who call this place "home." The Universe is *not* a zero-sum game; humans exist to help tip the balance in favor of Gift.

And cows are here to help us.

Index

Note: Page numbers in *italics* indicate illustrations.

P

paddocks
    accounting for new animals, 60–61
    during calving, 139
    for holistic grazing, *16*, 29, 31, 67, 68, 115, 116–17
    milk production influenced by, 131
    See also polytwine electrified fences
paint, for detecting heat, 151–53
parasites, internal and external, 168–69
pasture, defined, 115–16
    See also grass
pasture management. *See* holistic grazing
pasture wise animals, 6, 164–65
patrolling of fences, by cows, 68, 69
pens, for separating cows and calves, 145
perimeter fences, 67–70
permanent fences
    interior, 67, 71
    perimeter, 67–70
    See also fences
pharmaceuticals
    in creep feeds, 144–45
    evaluating fitness of the cow and, 48–49
pheromones, 150
photosynthesis, x, 13–14
pin bones, *46*, 126
placenta (afterbirth)
    retained, 140
    as sign of birth, 137
polled cows, 51, 61, 155
Polyface Farm, 28
polytwine electrified fences
    daily moving of, 31, 32, 68, 151
    for moving cows, 66
    prior exposure to, 49
    requirements for, 68–70
    turning off for newborn calves, 139

when adding in a new cow, 60, 77
Poppy (cow), 89
positional bloat, treating, 170–71
post-milking protocol, 94–95, *95*
pregnancy
    pre-calving changes, 126–27
    status when acquiring a cow, 50
    testing for, 157
pre-milking protocol, 89–90
prices of cows, 54–56
prior management, evaluating fitness of the cow and, 49–50
probiotics, in milk, 96, 105–6
Project Regenerate the Land, 122–23

Q

quality control, initial stripping for, 90
quality of life, 36, 37
Queenie (cow), 158

R

rags, for cleaning the udder, 74, 89
raw milk
    probiotic characteristics of, 96, 105–6
    safety considerations, 104
    See also milk handling
receptive (standing) behavior, 150, 151
recordkeeping, 94–95, *95*, 98–99
refrigeration of milk, 108, 109
regenerative grazing. *See* holistic grazing
retained placenta, 140
ringworm, treating, 169
Roberts, Heather, 62–63
Roberts, Mike, 62–63
rotational grazing, defined, 115
    See also holistic grazing
rumen
    grain effects on acidity of, 119

rumen (*continued*)
　physiology of, 15–17
　size in grass-fed vs. grain-fed cows,
　　114, 144–45
ruminants
　conversion of cellulose by, x, 15–17
　types of, 18–20
　See also family cows
Russell, Carlee, 147–48
Russell, Daniel, 147–48

S
safety concerns
　escaping cows, 65–67, 70–71,
　　77–78, 145
　milk handling, 103–9, *106, 107*
Salad Bar Beef (Salatin), 29
Salatin, Joel, 6, 29, 173
sanitation considerations
　milk handling, 103–9, *106, 107*
　milk processing equipment, 108
　teat dips, 95–96
　udder washing, 74, 89–90
Savory, Allan, 29
schedules for milking
　calf sharing and, 142
　factors in choosing, 88–89,
　　100–101
　first four days post-calving, 127–29
　once-a-day vs. twice-a-day, 100, 128,
　　132
Scottish Highlands cows, 44, 52
scours, 140, 144
seasonality of food, 14, 132
security needs, 77–78
　See also fences
sellers
　assistance from, 55, 56
　personal dynamics and, 54
semen, sexed vs. unsexed, 154
separating cow and calf, 145

settling (conceiving), 50
settling in, 65–78
　being prepared, 65–67
　comfort needs, 76–77
　infrastructure needs, 67–73, *69, 72*
　security needs, 77–78
　tools, 73–75, *75*
sexed vs. unsexed semen, 154
shared cows, 35
shelter needs, 71–73, *72*
　See also barns
shopping venues, 44–45
　See also acquiring a cow
single cows vs. herds, 57–61, *59*
size, evaluating fitness of the cow and,
　51–52
skimming of cream, 107, *107*, 109
Slan Abhaile farm, 10–11
slipping a calf (miscarriage), 157
Smith, Jay, 110–11
soil fertility
　cycle of restoring, 17–18
　intensive grazing benefits, 2, 4–5, 6
soiling, defined, 114
solar energy, photosynthesis and, x,
　13–14
spring houses, 109
staggers (magnesium deficiency), 173
stainless steel buckets, 74, *75*
stainless steel milk cans, 74, *75*
stainless steel storage containers, 107
stanchions, for milking
　factors in choosing, 81–86, *83*
　importance of, 72, *72*, 74
　training the cow to, 84–86, *85*, 127
standing (receptive) behavior, 150, 151
standing heat, 149, 150
Stephens, Jennifer, 135–36
Stephens, Tobias, 135–36
stickers, for detecting heat, 151–53,
　152, 157

About the Authors

SUZY HOLLING

Shawn and Beth Dougherty have been homesteading together since 1985. For the last thirty years, they've been on their farm, the Sow's Ear, in northern Appalachia, where they raise pigs, sheep, poultry, and cows for dairy and beef, with few-to-no inputs of feed, food, or fertilizer. Their ongoing project is the rediscovery of management patterns by which human beings generate ecological health and diversity simultaneously with food for people. The couple teach, write, and speak on inputs-free regenerative farming and the centrality of grass and dairy in sustainable food production. They are the authors of *The Independent Farmstead* (Chelsea Green Publishing) and the parents of eight children. Shawn and Beth are featured in dozens of podcasts and videos on regenerative farming, and they can be found at onecowrevolution.com.

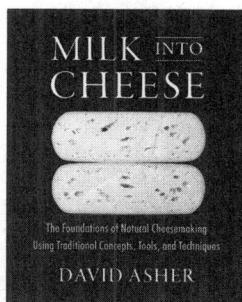

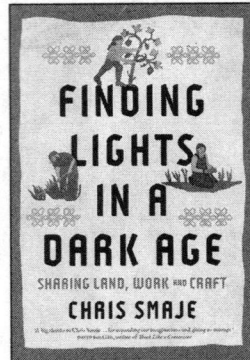

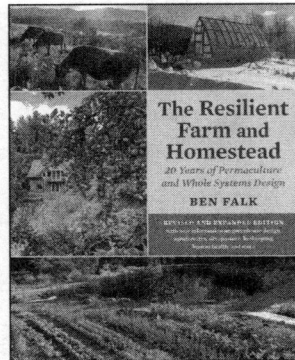